10/8/93

Rock fractures

Rock fractures

S. N. Chernyshev
Professor of Geology, Moscow Civil Engineering Institute

W. R. Dearman
Emeritus Professor of Engineering Geology, University of Newcastle upon Tyne; Visiting Professor of Engineering Geology, Polytechnic South West, Plymouth

Translated from the Russian by
G. B. Mirovsky

Butterworth–Heinemann
London Boston Singapore Sydney Toronto Wellington

 PART OF REED INTERNATIONAL P.L.C.

First published 1991

© Butterworth–Heinemann Ltd, 1991

British Library Cataloguing in Publication Data

Chernyshev, S. N.
Rock fractures.
1. Rocks. Fracture
I. Title II. Dearman, W. R.
551.87

ISBN 0–7506–1017–4

Library of Congress Cataloging-in-Publication Data

Chernyshev, S. N. (Sergeĭ Nikolaevich)
Rock fractures / S. N. Chernysev and W. R. Dearman; translated from the Russian by G. B. Mirovsky.
p.   cm.
Includes bibliographical references.
ISBN 0–7506–1017–4
1. Rocks—Fracture. 2. Rock mechanics. I. Dearman, W. R.
II. Title.
TA706.C46 1991
624.1′5132—dc20
90–40-692
CIP

Photoset by TecSet Ltd
Printed and bound by Hartnolls Ltd, Bodmin Cornwall

Contents

Introduction

Rock fractures form complex spatial systems. The structural pattern of fracture systems defines deformation, strength and permeability properties of rock masses and stratified formations, and also the distribution and methods for the development of a large number of mineral deposits. This book aims to show the rules governing fracture system patterns, ascertaining the sequence of and interrelationship between geological processes responsible for them, and exploring the possibility of using the structural parameters of fracture sets and systems to solve practical problems.

The concepts of original, tectonic, hypergene and other fractures are related to the hypothesis of the inherited evolution of rock fractures on the basis of the theory of inherited evolution of geological structures. This hypothesis agrees with the present theory of cracks in mechanics and with the ample factual data given in this book. It provides the basis for forecasting technogenic processes which occur in a rock mass under development, for example during mining or civil engineering.

Fractures originate, exist and evolve in rocks throughout their geological history. Fractures are an integral part of rocks. Fracture-system patterns often originate before a magma or a sediment is lithified. Bedding planes in sedimentary rocks are the first elements of a fracture system which define fracture arrangement. In igneous rocks the first fractures dissect a melt which is still mobile and partially crystallized. This is evidenced by the orientation of fractures relative to flow structures and their infilling with late magma differentiates. In effusives the first fractures are formed in freshly solidified lava on the surface of the still mobile flow. Specific fractures of metamorphic rocks, that is cleavage, originate during rock restructuring under metamorphism.

Originating at early stages of litho- or petrogenesis, fractures are present in rocks throughout their history. Their traces can be found even in eluvium where it is only by the fracture pattern that one can relate this new formation to a parent rock. A system of fractures in eluvium produced by weathering may have strong minerals resistant to weathering. For instance, quartz veins survive in granite eluvium with the parent rock transformed to sand, gruss or even clay. Calcite or limonite veins may be visible on the weathered rock surface. Yet eluvium also has fractures of another type. The same granite exhibits fractures along which feldspars were transformed into kaolin long ago. Quartz veins strengthen and reinforce a rock mass, whereas kaolinization zones make it weak.

As has been pointed out, fractures evolve together with rocks in which they are found. Studies of fractures help to reconstruct the geological history of a rock mass, mainly involving changes in rock mass stresses and deformations. Therefore, fractures have been the subject of exploration in tectonics and structural geology for more than 100 years. Investigation of non-tectonic fractures has lagged behind. So far there is no unified theory of the evolution of tectonic and non-tectonic fractures. The hypothesis on the inherited evolution of fractures proposed in this book is based on the general theory of inherited evolution of geological structures formulated by Peive (1956) and Yanshin (1951).

Information on rock fractures is of great practical value. Fractures serve as a route for movement of groundwater, oil and gas enclosed in pores and other rock interstices. Ore deposits are associated with fractures. Such fracture parameters as orientation, size and spacing define the process of rock mass disintegration due to weathering, karstification, and tectonic deformations. The character of disintegration of a rock mass during construction is determined by fractures, so it is only by knowing the main parameters of fracture sets that one can control this process. At present, high priority is given to methods for calculating strength, deformability, and water permeability of fractured rock masses. Hence, construction of a structural (geometrical) model of a fracture system is of great importance.

Current advances were made in studies of fracturing in the early twentieth century. To analyse the origin of fracturing, use was made of models (Levinson-Lessing, 1940) and of detailed field studies followed by processing of measurements of the orientation of fractures on either Schmidt or Wolff spherical projection nets. The same years saw new classifications of fractures, among which those of H. Cloos (1923) and Sander (1930) were the most important.

In the 1930s–1950s many Soviet geologists published papers on rock fractures which carried ample factual data enabling detailed description of different genetic types of rock fractures. An attempt was made at the time to account for rock fractures by the theory of deformation and failure taken from mechanics within the framework of the theory of formation of tectonic faults. The 1960s saw the application of the theory of mathema-tical statistics to the interpretation of ample field data, as well as of the Griffith's 1921, 1924 fracture theory to the explanation of rock fracturing (Pavlova, 1970). This trend called for fractures to be investigated directly on outcrops. At the same time another trend originated in the 1920s. Geologists sought to assess fracturing indirectly, based on such factors as the results of water injection into a hole, absorption of drilling fluid, and core recovery. Later, geophysical methods came into use for these purposes. None the less, indirect methods for studying fractures cannot be employed in all cases, although they should be studied through mechanics rather than statistics.

The first attempt to deal with rock mass properties on the basis of mechanics was apparently made by Lomize (1951) as applied to seepage problems. Later, Müller (1963), Romm (1966), Louis (1968), Snow (1968), Hoek (1973), Ruppeneit (1975), Chernyshev (1976, 1984), Wittke (1984). Jueguang (1986), followed suit in solving problems of deformation, strength and seepage. Yet the ideas expressed by the above authors were related to practice on a small scale. A major drawback was that their models of the geological medium were too abstract.

Studies of physical and mechanical properties of rock materials are far ahead of those of rock mass properties. There are numerous papers containing not only general methods for determining rock properties, but also actual lists of properties of a large number of rocks and minerals. A rock register (Melnikov *et al.*, 1975) has been published in the USSR, summarizing data on physical properties and technological parameters of rocks as an object of mining and processing; and similar compilations are available in the West.

Studies of rock properties are well advanced, suggesting that the parameters of rock materials composing a rock mass can be determined to a fairly high degree of accuracy and reliability. Moreover, in some cases values may be taken from reference books.

Thus, geological investigations of fracturing and studies of the mechanical properties of rocks have created the prerequisites for the development of a new trend in assessing rock mass properties. This trend is shown in this book on the basis of a combination of geological and geomechanical studies. In this respect, much use is to be made of computers in modelling complex geological structures. This will make it possible to discard costly geomechanical experiments in situ in favour of studies of rock samples with subsequent extrapolation of results on the basis of a model of rock mass geology.

A variety of classifications has been developed using simple engineering geological parameters of the rock material and the rock mass. These classifications have been applied in particular to the design of underground openings, including tunnels and large caverns. Bieniawski's (1976, 1979), 'Geomechanics Classification', and the Q-system developed by Barton, Lien and Lunde (1974), for example, were developed from earlier classifications, and have themselves been repeatedly refined (Hoek and Brown, 1980).

Part I
Rock fracture patterns

Chapter 1
Terminology and classification

1.1 Fractures

All rocks are affected by fractures which intersect and give rise to spatial sets. A total combination of fractures is referred to as fracturing. There are dozens of classifications of fractures and fracture sets and systems as to form, size, origin, and indirect evidence of degree of fracturing. Many authors (Zhilenkov, 1975; Krasilova, 1979) have suggested new classifications, which indicates that there are no logically substantiated and generally accepted classifications. Accordingly, the authors have attempted, while avoiding the introduction of new classifications and terminology, to refine existing concepts.

Definition

A fracture may be defined in structural geology as a plane of rupture along which there is little if any displacement (Anon, 1964), and yet displacement is implied if 'fracture' is held to include cracks, joints and faults (Bates and Jackson, 1980). This definition, however, seems inadequate as applied to many branches of present-day geology such as engineering, mining, petroleum geology, hydrogeology. Alternatively, a fracture is sometimes referred to as a complex-shaped cavity filled with gas, liquid or solid mineral matter. Fractures differ from other rock cavities, such as pores and solution cavities in that their extent along the walls in all directions, conventionally termed 'length', far exceeds the distance between the walls, referred to as 'width' or 'aperture'. The definitions of a fracture, both as a surface or plane and as a three-dimensional body, are in agreement with each other since they characterize it from different aspects. Speaking in favour of these two definitions, Kosygin distinguishes two aspects in describing fractures. In his opinion, they can be regarded, on the one hand, as disjunctive boundaries and, on the other, as 'geological bodies whose thickness is much smaller than their length which were formed as a result of different processes (mineralization, metamorphism, crushing, brecciation, mylonitization, metasomatic replacement, etc.) related to a particular boundary' (1974, p.117).

If it were not for the second, genetic part, this definition would be inadequate since by a purely morphological definition a fracture is somewhat similar to an interlayer or a layer. This is quite reasonable from the standpoint of geomechanics inasmuch as an interlayer, in the same way as a fracture, serves as a surface of weakness, and a fracture with permeable filling does not differ hydraulically from an interlayer of variable thickness.

Fracture classification

The form of a fracture reflects its evolution. Through geological time it can be a tension fracture, a shear or slip plane, a filter channel with eroded walls or a place where matter transported by groundwater has accumulated. A fracture is characterized by inherited evolution, which makes it fairly difficult to classify it genetically, i.e. to relate a fracture to a particular process (contraction, weathering) responsible for it. That is why the genetic classification, far from being impartial, has a preference for some particular process. Yet, the genetic terminology and classification should in no way be disregarded since these reflect great experience in studying rock fractures. The temptation should be avoided to describe fractures purely geometrically, as required in geomechanics. Otherwise, we shall not be able to extrapolate and interpolate observational data.

Geometry

The geometrical parameters of rock fractures can be determined to a high degree of accuracy, and hence classification of fractures as to their form and size is fully impartial. The main distinguishing feature of a rock fracture is, as noted above, that its length greatly exceeds its width. Therefore, in tectonics, while the trace of a fracture on any surface of exposure is a line, a fracture proper is shown as a plane. In engineering geology a fracture cannot be shown in that way. It should, rather, be illustrated as segments of two parallel planes, each plane approximating to one of its rough walls. The distance between these planes, referred to as 'fracture width', is denoted by b.

It should be noted that in Western terminology a distinction is made between 'fracture width' and 'fracture aperture'; 'aperture' is the perpendicular distance separating the adjacent walls of an open fracture (Brown, 1981, p.35), whereas the term 'width' refers to the distance between the walls of a filled fracture.

The length of a portion of the plane termed 'fracture length' is denoted by l. The rough walls of a fracture come into contact with each other, and the relative area of rock contacts, i.e. the ratio of the contact area to the total area of the fracture wall, is denoted by s. Curvature and roughness of a fracture wall, i.e. characteristics of its deviation from the ideal plane, have to be taken into account in solving some engineering problems.

Apart from fracture configuration, its spatial attitude and position are also essential factors. Fracture orientation is determined by dip azimuth α and dip angle β. It is also necessary to know the distance of the fracture from any particular object, say from a neighbouring parallel fracture. The fracture position relative to neighbouring fractures, structural features (fault, fold) or artificial structures (axis of a dam) can be shown graphically, for example on a plan. The following definitions are generally accepted:

- Fracture set: a group of fractures which run more or less parallel to each other.
- Fracture system: two or more fracture sets which intersect at a more or less constant angle.

A system generally includes fractures of different genetic types. In addition to geometrical parameters, a system of fractures is characterized by their particular mutual distribution (angles, distance between parallel fractures), as well as by the extent to which the rock mass is dissected into joint blocks. Depending on fracture length and spacing, the joint blocks are entirely separated from each other or somewhat interconnected through rock bridges. A system of fractures bears a great resemblance to a crystal lattice whose crystals display eminent, perfect or imperfect cleavage.

A combination of fractures featuring similar origin and development is attributed to one and the same genetic type. The similarity covers habit, i.e. roughness and curvature of fracture walls, fracture terminations, etc. The genetic type generally includes several fracture sets.

Age of fractures

The age of fractures should be determined both for geological analysis and for constructing geometrical models of fracture systems to solve some particular engineering problem. A distinction should be made between the geological and the relative age of fractures. The geological age of fractures, just as that of rocks, can be determined by an absolute or relative geochronological time scale. To indicate geological age, one may refer to some particular events in the history of a rock mass which occurred at the time the fractures were formed. For instance, when the fractures are termed 'lithogenetic' or 'weathering', their age is determined to some degree by that of the rocks or of the relief of the top of bedrock. The term 'relative age' is used to distinguish between the age of fractures in a particular rock mass. Fractures of one and the same geological age and genetic type can be subdivided into several generations by relative age. The relative age of fractures is determined by a number of formal factors, irrespective of their geological age.

The rock mass

Rock material is a rock with rigid crystallization or intergranular cementation bonds. A rock mass, a combination of rock materials, is a complex geological body comprising a variety of rocks featuring both a common origin and subsequent evolution. Its boundary is defined by geological and genetic factors, for example igneous rocks occurring on one side of a boundary and sedimentary rocks on the other, or metamorphosed Palaeozoic marine deposits on one side and Cainozoic loose continental deposits on the other. The rock mass boundary may be composite. Not only rocks with rigid bonds are involved. Each rock mass is characterized by a particular lithologico-petrographic composition, mode of occurrence and fracturing pattern. In engineering–geological studies, one should have adequate information on the strained state of a particular rock mass, its temperature and moisture content.

Fracture filling

Fracture filling is a loose soil or rock formed in a fracture. It has a particular chemical composition, structure, and texture. In spite of the fact that fractures are fairly narrow, some of them have several layers of filling of different age and origin. Filling should be considered as a rock or soil formed under specific conditions within the inner space of a fracture. Little progress has been so far made in this respect in engineering geology, although the importance of the mineralogical, grain size and other characteristics of fillers is fully acknowledged.

Discontinuity

In the West, the collective term for all types of joints, weak bedding planes, weak schistosity planes, weakness zones, shear planes, faults and contacts, in other words all 'breaks' in rock masses, is 'discontinuity' rather than 'fracture'. Discontinuity, therefore, is a general term for any mechanical break in a rock mass, and incidentally in a soil mass, having zero or low tensile strength.

In the following discussion on rock mass fracture (Section 1.4), the term 'discontinuity' of fracture sets originally used by the Russian author has been substituted by 'persistence'. Persistence implies the areal extent or size of a fracture within a plane. This term is in line with the recommendations for the suggested methods of rock characterization published by the International Society for Rock Mechanics (ISRM) (Brown, 1981), and acceptance of its use internationally will help to avoid misconceptions and possible misunderstanding.

1.2   Fracture form and the structure of fracture systems

Fracture form

Fractures are in fact rock mass discontinuities due to brittle failure. In the quasi-continuous and quasi-homogeneous part of a rock mass, fractures run along the plane reflecting the constant spatial position of axes of stress and the mirror symmetry of stress fields in their flanks. A fluctuation of stress due to rock heterogeneity or proximity to the layer boundary causes fracture walls to deviate from the ideal plane, thus imparting to each specific fracture an individual appearance. There are tension and shear fractures (Figures 1.1 and 1.2) as regards wall shape. The former have a rough surface with some rounded portions. Surface roughness is apparently determined by rock grain size, less than a millimetre to a few centimetres in fine- and coarse-grained rocks respectively. Shear fractures, which may be slickensided, feature smaller deviations from the symmetry plane, their surface following a regular step-like pattern. The step surfaces are inclined to the fracture symmetry plane at an angle of 5–15°. They are cut off by scarps which are almost square to the step plane. Tension fractures develop in the plane of principal normal stresses, whereas shear fractures follow the direction of major tangential stresses.

(a)

(b)

Figure 1.1 Surfaces of shear (a) and tension (b) fractures in medium-grained granite, Kolyma Hydro foundation, rule 20 cm long

Figure 1.2 Surfaces of shear fractures (smooth, inclined to the right and left) and tension fractures (rough) in Jurassic clay shales, Kolyma valley, rule 20 cm long

However, tension and shear fractures are not the only ones found in rocks. There are also fractures which have been ground by the slipping process with well-polished furrows or slickensides running along their walls in the slip direction. These fractures have a great role to play in governing strength, permeability and other properties of a rock mass. While drastically reducing its strength, they may increase or decrease water permeability. It is, therefore, advisable to single them out as a separate morphological type. The slip fractures may be metamorphosed shear or tension fractures. There is another morphologically and genetically isolated type of surface of weakness in a rock mass, namely the bedding surface whose roughness is due to the irregular granulometric composition of a sediment, ripple marks, bottom structures, etc. Thus four types of surface of weakness can be distinguished according to the shape of the fracture walls, namely: (i) bedding surface; (ii) tension fracture; (iii) shear fracture; (iv) slip fracture.

Fracture systems

The structure of fracture systems varies greatly in a rock mass depending on lithological, petrographic and tectonic factors. For a small rock mass where the system of fractures can be taken as spatially uniform, its structure is most prominently expressed by the configuration of a joint block. There are five types of jointing: (i) spheroidal; (ii) platy; (iii) columnar; (iv) parallelepipedal; (v) polyhedral. Apparently the empi-

Table 1.1 Geometrical classification of fracture systems

Fracture system description	Graphic representation	Classification basis	
		Stressed state of rock mass responsible for fractures	Rock mass anisotropy due to fracturing
I Spheroidal		$\sigma_1 = \sigma_2 = \sigma_3$	Isotropic
II Polygonal axisymmetric		$\sigma_1 = \sigma_2 > \sigma_3$ or $\sigma_1 > \sigma_2 = \sigma_3$	Transverse–isotropic
III Regular equiangular		$\sigma_1 > \sigma_2 > \sigma_3$	Anisotropic
IV Chaotic, asymmetric		Stressed state changes with time	Isotropic

rical approach has not led to a unified classification, so that here a theoretical approach to classifying fracture sets is proposed

The environment of fracture formation is generally expressed by stress tensor. All three possible combinations of normal stresses are realized in a rock mass, namely: (i) $\sigma_1 = \sigma_2 = \sigma_3$, with all three principal normal stresses equal;[*] (ii) $\sigma_1 = \sigma_2 > \sigma_3$ or $\sigma_1 > \sigma_2 = \sigma_3$, with two out of three principal normal stresses equal; (iii) $\sigma_1 > \sigma_2 > \sigma_3$, all three principal normal stresses are different. Three types of stresses are realized in three types of fracture systems provided stresses are constant both with space and time. Breaking stresses in a material changing with time provide a fourth type of stressed state responsible for a particular system of fractures. Each of the four systems is characterized by its own symmetry group (see Table 1.1).

In the first case the spherical stress tensor results in the development of tension fractures in the form of concentric spheres and radial division planes. The symmetry group of such a figure is known to have a centre of symmetry, an infinitely large number of symmetry axes of infinite order, and an infinitely large number of symmetry planes. Fracture systems of this type, with statistical departures from the ideal form, may be observed in spherulitic lavas, a zone of weathering, and as a result of blasting. Together with related jointing they are called spheroidal, globular, or shell-like (Figure 1.3(a)).

The second combination of principal normal stresses with two out of three stresses being equal can be visualized as an ellipsoid of revolution, provided all the three stresses are of the same sign, which fails to cover all the versions. In this case the stresses have a centre of symmetry, one symmetry axis of infinite order, an infinitely large number of two-fold rotation axes, and an infinitely large number of symmetry planes, one of which is normal to the symmetry axis of infinite order and others intersect one another along this axis.

The above automorphic[†] transformation make up a symmetry group which is fixed on rock disintegration and is preserved in the geometrical form of blocks and fractures once stresses are relieved. The stress field in question originates and is statistically realized in columnar joints of volcanic rocks (Figure 1.3(b)), desiccation fissures in sedimentary strata (Figure 1.3(c)) and in the structure of fractures in salt domes. In all these cases fractures develop along the element of a cylinder with σ_1 serving as its axis. These fracture systems develop under tensile-stressed conditions when the equal minor principal normal stresses are of negative sign. Under conditions of compression shear fractures are formed to follow the generator of cone whose axis coincides with axis σ_1.

The third type of the stressed state ($\sigma_1 > \sigma_2 > \sigma_3$) can be visualized as a general ellipsoid, all the three principal normal stresses being of the same sign. With rocks disintegrated by these stresses, systems of tension and

---

[*] Fractures are caused by tensile stresses.

[†] The term 'automorphic', a synonym of 'euhedral', refers to the shape of a crystal that is completely bounded by its own 'rational faces'; such crystal faces usually have low Miller indices, and a resultant simple crystal form. By analogy, joint-bounded blocks will have the simplest form consistent with the symmetry of the generating principal normal stresses.

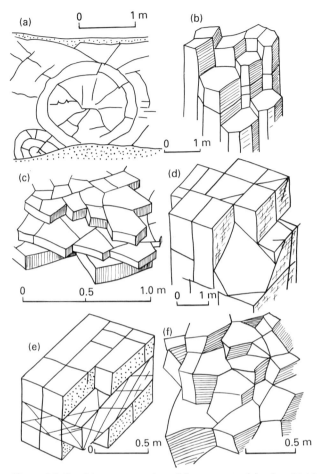

Figure 1.3 Graphic representation of fracture sets: (a) spheroidal in andesites; (b) polygonal in basalts; (c) polygonal in limestones; (d) system in granites; (e) system in sandstone and mudstone; (f) chaotic in dolerite

shear fractures are formed to bound prismatic blocks. The blocks have the following symmetry group: a centre of symmetry, a finite number of symmetry axes of finite order, and a finite number of symmetry planes. This third type of the stressed state is typical of stress fields on longitudinal compression joints and at faults. With inequality $\sigma_1 > \sigma_2 > \sigma_3$ the most common fracture system (Figure 1.3 (d,e)) is formed.

Should the generations of fractures be superimposed in succession on each other with no healing, asymmetric blocks of the fourth type are formed on the principle that the symmetry of the field transforming the body is superimposed on the body symmetry already formed. The resultant shape of the body retains only those elements of its symmetry which coincide with the superimposed field symmetry elements (Shafranovsky, 1968). With stress fields superimposed several times, there are no common symmetry elements. The blocks become asymmetric. This process is observed in crush zones, at faults, at intrusive contacts (Figure 1.3 (f)), in a zone of weathering, i.e. in most mobile portions of the Earth's crust.

The four types of jointing based on the analysis of relations between the principal normal stresses generally agree with jointing types singled out empirically (Rats and Chernyshev, 1970). Each type has its own symmetry group and fracture system. The groups differ as follows: I – infinitely large numbers of symmetry elements of infinite order; II – finite number of symmetry elements of infinite order and infinitely large number of symmetry elements of finite order; III – finite number of symmetry elements of finite order; IV – no symmetry elements.

In addition to automorphic transformations of individual blocks (reflection, rotation), systems of fractures enable transfers along particular axes with a fixed spacing. The number of transfer axes is determined by the block symmetry group. For the spheroidal set with an unsystematized arrangement of blocks the number of transfer axes is infinitely large and they are located in all directions. The transfer spacing is equal to the average diameter of a block with concentric shell-like jointing. The coincidence of the structure as a result of transfer and superposition is possible only in a statistical sense, just as the other automorphic transformations mentioned above. The transfer is feasible only within a lithologically uniform and structurally isolated block of a rock mass with a statistically uniform stress field. Such geological areas are observed in fractured rock masses where the statistical functions of distribution of fracturing parameters are constant and hence can be determined. Transfer in conjunction with rotation is possible when moving from one portion of a rock mass to another with a similar fracture pattern, say, from one fold limb to the other. This case can be regarded as curvilinear symmetry. Similarity symmetry is observed with the fracture frequency and width trend at faults and in the hypergenesis zone.

Up until now it has been apparent that the configuration of fracture systems is largely determined by stress fields without reference to rock mass composition. However, rocks composing a particular rock mass have an influence on fracture symmetry via a stress tensor which is equally dependent on a source of stress and on rock mass composition. The location of axes of principal normal stresses is adjusted by structural rock features. The transfer spacing is determined by rock strength with equal stress intensity. The relationship between fracture systems and rock composition shows up vividly in a stratified formation consisting of different rocks. The fracture systems in, say, sandstone and mudstone are different in one and the same formation (Figure 1.3(e)).

The major morphological types of fracture systems are clearly shown in circular diagrams (Figure 1.4). There is no difference between circles 1 and 4. The diagrams are really identical, but spheroidal and chaotic fracture patterns are quite different. The former has curvilinear and closed fractures, whereas the latter has mostly rectilinear fractures. In the first case, the fracture system develops in the presence of an isotropic stress field constant with time. In the second case, stress isotrophy is attained only within a considerable period of time.

The practical engineering–geological importance of the proposed classification, as the authors see it, is as follows. It covers all the combinations of

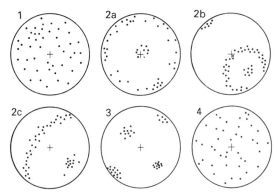

Figure 1.4 Schematic representation of fracture sets in circular fracturing diagrams:
1 – spheroidal; 2 – polygonal (a, b, c); 3 – regular; 4 – chaotic

fractures as regards the pattern of their arrangement. When studying properties of a rock mass as a continuum, each of the above-mentioned fracture systems usually generates some particular rock mass anisotropy. The spheroidal system corresponds to the isotropic medium; the polygonal system to the transverse-isotropic medium; the regular system to the orthotropic or complex-anisotropic medium; and the chaotic system to the isotropic medium.

Each type of fracture system has its own model of a rock mass when passing from properties of discrete blocks and fractures to properties of a rock mass as a continuum. For instance, the model of the tensor permeability theory has been developed for the regular equiangular system of fractures. The model of the linear element theory corresponds to the axisymmetric polygonal system and regular equiangular system of fractures. Slope stability calculation is performed for the equiangular and axisymmetric polygonal system of fractures. Fracture system symmetry provides the basis for determining fold parameters and fracture discontinuities. With one infinite-fold symmetry axis in the system of feather or Riedel (1929) fractures in a fault, the arrangement of the fault and flank displacement can be identified. The classification enables calculated models and natural situations to be identified and hence reveals those natural situations for which no methods have yet been developed. The classification concludes the genetic analysis of static geological space. It provides the basis for calculated models of a rock mass described in Part II of the book.

1.3 Fracture size

Length of fractures

The length of fractures dealt with in geology varies over a wide range from 10^{-4} to 10^{8} cm. There are 4–5 classes (Table 1.2), 10^{-3}, 10 and 10^{4} cm being regarded by many authors as boundaries between the classes. The

Table 1.2 Fracture and fault levels classified by absolute length

Fracture length (cm)	After Muller (1974)	After Rats and Chernyshev (1970)	After Zhilenkov (1975)	After Krasilova (1979)	Recommended
10^8	Faults	Large tectonic faults	Faults of the 1st and 2nd order		Large tectonic faults
10^7					
10^6				Megafractures – faults	
10^5		Faults	Faults of the 3rd order		Faults
10^4					
10^3	Gigantic	Macrofractures or fractures	Large fractures	Macrofractures and mesofractures	Fractures — Long
10^2	Large		Fractures		Fractures — Medium
10	Small	Microfractures		Microfractures	Fractures — Short
1	Blind fractures				Microfractures
10^{-1}					
10^{-2}					
10^{-3}		Crystal lattice defects		Ultrafissures	
10^{-4}					

Levels

medium class comprises what Müller (1963), Rats and Chernyshev (1970) and Zhilenkov (1975) call fractures. It is this class, as is evident from the title of the book, that receives primary consideration. Microfissures and faults are dealt with in less detail.

Rock fracture generally develops from dislocation (Figure 1.5), which is a deformed crystalline structure covering only a fairly small number of atoms at each particular moment and capable of migrating under certain conditions with the crystal recovering its strength on the dislocation path. In migrating, two neighbouring layers of the crystal lattice slip along each other, the structural defect of the lattice moving from one site to another. If dislocation originating at one crystal face runs through the whole crystal to reach its other face, one part of the crystal shifts with respect to the other along the slip plane by the interstitial spacing. A 'step' is formed at each crystal face, the crystal strength remaining unaffected. Such deformation is termed 'plastic'. Dislocation may also migrate by creeping from one lattice plane to another.

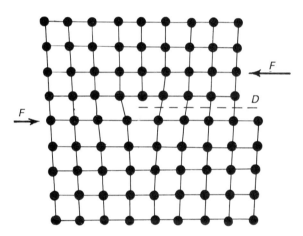

Figure 1.5 Dislocation D resulting in crystal lattice from displacement under the action of force F.

Dislocations develop during the growth of any real crystal. Their density generally ranges from 10^2 to 10^8 per cm^2. With the crystal subjected to such action as cold rolling, and high-frequency vibration, dislocations become more dense with almost the whole energy absorbed by the crystal concentrated therein.

The brittle failure of crystals under stress takes place on local defects of the crystal surface. First, dislocations accumulate, then they merge, which coupled with increasing stress gives rise to microfractures.

With a few dislocations merged, a cavity is formed inside the crystal to prevent further slip of dislocations in the crystal lattice. After cessation of slip, plastic deformation of crystals discontinues and their brittle failure begins. Thus a fracture, even in infancy, is determined as a cavity. The brittle failure of crystals generally takes place along the cleavage surfaces coinciding with the direction of crystallographic lattice planes across which the chemical bonding force is minimum (Bravais–Wolf rule). A rupture along the cleavage plane does not take place over the entire crystal section at one time, but rather by merging infant fractures in different portions of the crystal. In this case there occurs a changeover from one cleavage plane to another running parallel, which accounts for a step-like pattern of microfractures in crystals.

Jointing is another form of potential surface of weakness in crystals. Joint planes, as with cleavage planes, are not fractures until dislocations merge. The location of joint planes is determined not by weak binding in an ideal crystal lattice as in the case of cleavage, but by the pattern of its distortions. The latter may be due to double interlayers, regular distribution of isomorphic impurities or submicroscopic inclusions.

In solving engineering problems the relation between fracture length l and rock mass area L affected by a structure is of great significance. Tentatively, there are three levels: (i) $l < L/50$; (ii) $L/50 < l < L$; (iii) $l > L$. Thus, the practical value of the classes is different. Dislocations occur in rock grains, while microfractures develop in grains and small crystalline aggregates. Both of them may be found in the grains of a non-cohesive soil to determine their frost resistance and abrasion. Microfractures determine tensile and compression strength of rocks and other properties in small volumes (a few cm^3 or dm^3). Morphological studies of fractures in engineering geology are mostly of scientific value. Fractures (macrofractures) are longer than a sample but shorter than the base of most engineering structures. Their effect on the structure is assessed by statistical description of the fracture system and mechanical properties, the latter being investigated using field methods.

Faults are longer than the length of a rock mass area L. They are generally not numerous and are studied individually. It is interesting to note that there are not only tectonic faults. A fault is any rock mass discontinuity extending for more than 100 m, for example caused by landslides. The dimensions of major tectonic faults generally exceed those of engineering structures both in length and width. These are studied in engineering–geological zoning. The design model of the base of an engineering structure allows for faults as an individual element. Faults are often represented geologically by a group of fractures, including mylonitization and brecciated zones. Therefore, it is only conventionally that we can classify them together with fractures. However, faults resemble fractures on photographs taken by remote sensing and, vice versa, some structures look like faults upon magnification (Tchalenko, 1970).

Fracture width or aperture

The width of fractures studied in engineering geology varies from microns to a few metres, this range having been broken down into levels differently by various authors (Table 1.3). It is considered that permeability should be the main criterion in this respect. The velocity of flow in a fracture is known to increase as the square of its opening, and the rate of flow as the cube of its opening. In terms of water permeability a fracture 1.0 cm wide is equal to a thousand fractures 0.1 cm wide having other similar para-

Table 1.3 Fracture and fault levels classified by absolute width

Fault crush zone or fracture width (cm)	Levels			
	After Romm (1966)	After Farran and Thenoz (1965) in Jaeger, 1972	After Zhilenkov (1975)	Recommended
10^4 10^3 10^2	Macrofractures		Large fractures	Fault crush zones
10		Macrofractures		Fissures (gaping or filled)
1				Wide — Fractures
10^{-1}			Fractures	Medium — Fractures
10^{-2}				Narrow — Fractures
10^{-3}	Microfractures	Microfractures		
10^{-4}			Small fractures	Capillaries
10^{-5}		Microfissures		Subcapillaries

Fractures

meters. Fracture width changing by an order of magnitude is, therefore, an essential feature of the classification. The classes are named differently as compared to those in Table 1.2, which enables such characteristics as 'short, but wide fractures', 'gaping faults', 'short and narrow fractures' to be defined. These characteristics are important inasmuch as fracture length and opening are for the most part inadequately correlated.

The class boundaries indicated in Table 1.3, as those in Table 1.2, separate fractures which are of essentially different practical importance. Subcapillaries (opening less than 1 micron) are fractures whose width is of the order of the dimensions of pores in clayey soil. Water movement in these fractures is impeded by bound water. However, it is not by any means eliminated and conforms to the Bousinesque law in the case of fractures wider than 0.2 micron (Romm, 1966). Hydrostatic pressure can be transmitted through subcapillaries. The distance between the walls of closed fractures healed in terms of seepage by bound water is about 1×10^{-6} cm.

Capillaries are fractures whose width is of the order of the dimensions of pores in a silt. Gravity water can freely move in such fractures. However, it is confined by capillary forces in the aeration zone, which greatly affects rock mass properties, in particular, strength. Water permeability of a rock mass with capillary fractures is extremely low and cement grouting is usually not undertaken in construction. Dewatering is not required either.

Opening of narrow fractures is of the order of dimensions of pores in sand, and that of wide fractures of pores in gravel. A rock mass with fractures of such width is permeable and is in need of grouting. It can be used as a water supply source. A variety of fractures makes it possible to represent the rock mass as a continuous water permeable medium. Seepage theory can be applied in this case. In terms of size, fractures can be correlated with pores between pebbles and boulders. They should be modelled individually as a permeable bed or free space.

1.4 Rock mass fracturing

The study of rock mass fracturing, usually with a fracture system of more than one set, involves the practical determination of the following:

(a) the size of joint-bounded blocks;
(b) the extent of the division of the rock mass into individual separate blocks; and
(c) the volume of void space resulting from the degree to which the fractures are open, that is, rock mass 'porosity'.

The first and third of these characteristics are similar to the granulometric composition and porosity of a soil. The second characteristic is of special importance in rock engineering in that it takes into account the presence or absence of rock bridges between joint blocks, involving estimation of the persistence or extent of a given fracture or set of fractures, and the combination of sets into a three-dimensional system in which the sets may, or may not, be mutually orthogonal. Persistence can play a significant role in determining the behaviour of a rock mass, particularly in studies of the stability of rock slopes and underground openings.

All three factors influence the hydrogeological properties of a rock mass, the movement of groundwater and its availability to wells, effective grouting of rocks in foundations, downward migration of underground water as a chemical weathering agent, the formation of ice-filled joints under permafrost conditions, and so on.

An ordered approach to rock fracturing conveniently starts with consideration of a single joint set, progressing to the areal properties of two sets, before considering the three-dimensional reality of the natural rock mass.

Persistence of fracture sets

The degree to which a rock mass is separated into discrete blocks is a most complex phenomenon. Yet the role of the persistence factor cannot be overemphasized in determining strength, water permeability and other properties of a rock mass. A distinction should be made between sets of discontinuities which are persistent, sub-persistent and non-persistent, as illustrated by Müller (1963) and Price (1966). Depending on the degree of persistence, rock masses may be described as:

(a) completely fractured, with the rock mass fully separated into blocks by persistent fracture sets;
(b) moderately fractured, with some blocks separated and others connected by intact rock bridges (Piteau, 1973; Robertson, 1971) by sub-persistent fracture sets; and
(c) solid, with the rock mass not separated into blocks, but some individual non-persistent fractures are present. Moderately fractured rock masses are the most common, although problems in hydrogeomechanics have been solved mainly for rock masses assumed to have continuous, persistent fracture sets.

It is fairly difficult to choose an appropriate parameter to characterize fracture system persistence since it is subject to scale effect and varies in different directions. Indeed, any fragmented, completely fractured rock mass in the weathered layer comprises individual small but solid blocks and, vice versa, any large solid rock mass, which, for example, can be found in the central portions of granite batholiths, is not boundless but has certain limits determined by tectonic faults. Thus, a fracture system can be classified in terms of persistence of fracture sets without indicating the parameters of the fracture sets or rock mass dimensions.

Quantitative parameters to assess persistence were proposed by Pacher (L. Müller, 1961) as well as by Rats and Chernyshev (1970) as a function of fracture spacing and length. According to Pacher, persistence is expressed as the areal extent or size of a fracture within a plane in the rock mass section; in other words there is a two-dimensional aspect to be considered. It is impossible, Müller believed (1961), to determine this reference value without breaking down the rock mass, although natural exposures and excavations with rock faces in more than one direction enable realistic estimations to be made.

Quantitative description of persistence

In rock exposures, an attempt should be made to measure fracture lengths in both the dip and strike directions, if the exposure has a suitable shape. It may be possible to plot size frequency histograms for each discontinuity set present, and the modal trace lengths can be described according to the following scheme:

Very low persistence	< 1 m
Low persistence	1–3 m
Medium persistence	3–10 m
High persistence	10–20 m
Very high persistence	> 20 m

Analyses of dip and strike lengths (Robertson, 1971) reveal that discontinuities tend to be approximately isotropic in dimensions. A discontinuity termininating in solid rock will then tend to be circular, whereas it will presumably be rectilinear when terminating against other discontinuities.

Assessment of persistence in a fracture system

The following criteria (Kolichko, 1966) can be applied quantitatively to assess a regular system of fractures. A persistent system of fractures is one with three approximately orthogonal sets among many others for which the inequalities $\bar{l}_1 > 2a_2$; $\bar{l}_1 > 2a_3$; $\bar{l}_2 > 2a_1$; $\bar{l}_2 > 2a_3$; $\bar{l}_3 > 2a_2$; $\bar{l}_3 > 2a_1$ hold true. A non-persistent system of fractures is the one where none of the above-mentioned inequalities holds true. Finally, a system of fractures for which only some of the above-mentioned inequalities hold true is termed sub-persistent.

Persistence of discontinuity traces intersecting a scanline

The preferred method, in Western countries, of estimating discontinuity spacing and trace length is by scanline survey (Priest and Hudson, 1981). A scanline survey involves sampling and measuring only those discontinuities that intersect a line set on the surface of the rock mass. In a tunnel, for example, the scanline will be set at a height convenient for surveying. As a method, it is a very suitable means of obtaining a sample, but in the case of trace lengths, the raw data contain an inherent and significant bias.

During a scanline survey, it is desirable to record quantitative information on the extent of the fractures intersected. In a planar, or nearly planar rock face, the simplest measure of fracture extent is the length of the trace produced by the intersection of a given fracture with the rock face. Priest and Hudson (1981) have discussed the estimation of trace length distributions. If the aim of the investigation is to determine mean trace discontinuity length, the mean value can be computed by measuring the trace length of N discontinuities intersected by the scanline. One drawback of this method is that the scanline preferentially intersects those discontinuities with a longer trace length, giving a biased sample (Figure 1.6(a)). The other difficulty is that large discontinuities may extend beyond the exposure, giving a number of trace lengths that are censored at some value

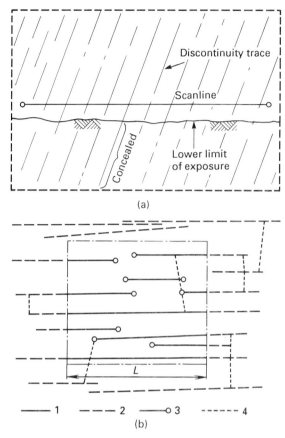

Figure 1.6 Method for determination of average fracture length in limited outcrop area: 1 – visible portion of measured fractures; 2 – same fractures outside outcrop; 3 – visible fracture terminations; 4 – fractures of another set

dependent on the size of the exposure. For practical purposes it is usually necessary to set the scanline near to the base of the rock face, effectively limiting measurements of trace length to that portion of the trace extending above the scanline. Once these effects have been recognized and allowed for in the analysis, it is a relatively simple matter to obtain a reasonable unbiased estimate of mean trace length termination frequency and hence of mean trace length (Priest and Hudson, 1981).

In the example illustrated in Figure 1.6(a), there is only one discontinuity set, and the fractures arise and terminate in intact rock. This is an artificial arrangement selected for the purpose of analysis, and one rarely found in nature.

Persistence of discontinuity traces in a planar area

An alternative approach involves a selected area of rock face in which fractures may extend beyond the exposure and may terminate either in intact rock or against a second, crossing fracture set (Figure 1.6(b)). Now under these conditions it is apparently expedient to assess persistence as $\bar{l}_1/\bar{a}_2$, where $\bar{l}_1$ is an average length of fracture in set 1 and $\bar{a}_2$ is an average fracture spacing in set 2 intersecting the first one. For instance, fractures of the first set (Figure 1.6(b)) are shown by lines 1 and 2 and those of the second set by line 4. The a value can be measured in outcrops and exploratory openings to a high degree of accuracy. The l value cannot as a rule be measured directly because many fractures, in terms of their length, go beyond the boundaries of the exposure. The following relation serves to determine the average length $\bar{l}$ of fractures in an outcrop:

$$\bar{l} = L[(2N/n) - 1] \ldots (1)$$

where

    L is the length of the measuring site;
    N is the total number of fractures within the site;
    n is the number of fracture terminations.

Example

As an example of the determination of average length, consider the 16 horizontal fractures shown in Figure 1.6(b). The outcrop area is square. The fractures outside the square are not visible. Assuming that the fractures are statistically homogeneous, their length can be determined by calculating the number of fractures in the limited outcrop area and the number of their terminations visible in the same outcrop area. Most fractures usually terminate beyond the outcrop boundaries, hence the number of their terminations is less than that of fractures proper. The number of terminations n is 9 (Figure 1.6(b)), and the number of fractures in the outcrop N is 10. Hence the average length of a horizontal fracture $\bar{l}$ is 5.8 m. The actual average length for the total combination of fractures is 5.1 m.

Fracture spacing

Two parameters of fracture frequency may be determined, namely:

(a) M_f – a modulus of fracturing – the number of fractures per 1 running metre in a specified direction in the rock mass (Kriger and Preobrazhensky, 1953; Belousov, 1962; Stini, from Müller, 1963; Smekhov, 1969; Anon., 1981a); and

(b) a – the fracture spacing – in a set measured in a direction perpendicular to the fractures (Müller, 1963; Rats and Chernysev, 1970; Jaeger, 1972; Richter, Molek and Reuter, 1976; Brown, 1981).

The a value is preferable as it is more convenient in statistical processing and geomechanics calculations.

    In Western studies, mean discontinuity (fracture) spacing values are obtained from scanline surveys (Priest and Hudson, 1976, 1981) on exposed rock surfaces, or by determinations on drill core. Where discontinuities occur in one or more sets, the discontinuity frequency along a scanline is a function of the orientation of the scanline (Terzaghi, 1965). There is considerable justification for assuming that spacing distribution of discontinuities in many rock masses is a negative exponential distribution with the mean spacing of discontinuities being the reciprocal of the average number of discontinuities per metre (Hudson and Priest, 1979; Wallis and King, 1980). This value can be calculated by dividing the number of scanline intersections by the total scanline length.

A commonly used quantitative measure of the fracture state of rock cores is Rock Quality Designation (RQD) proposed by Deere (1968). RQD is the percentage of rock recovered from a core run as sound lengths which are 100 mm or more in length. Only core lengths determined by natural fractures should be measured. If a scanline is regarded as analogous to a borehole, the general relation between RQD and $L(l)$ is:

$$RQD = 100[1 - L(0.1)]$$

where $L(l)$ is the cumulative length proportion, the proportion of scanline that consists of all spacing values up to a given value l, in this case 0.1 m, the RQD threshold value.

For the negative exponential distribution of spacing values:

$$RQD = 100e^{-0.1\lambda}(0.1\lambda + 1)$$

where λ is the mean discontinuity frequency, the average number of discontinuities per metre (Hudson and Priest, 1979).

Fracture spacing classification

Fracture spacing can be applied only to a regular system of fractures (Table 1.1); for chaotic systems a modulus of fracturing should be determined, although it provides only a rough classification characteristic of fracturing. Table 1.4 lists classifications of different authors, the number of fractures per 1 running metre being converted to fracture spacing.

There is a variety of fracturing classifications in terms of fracture spacing with division into 4–5 classes most common. As the fracture spacing increases, so does the classification interval length, which corresponds to the left-asymmetric log-normal distribution of fracture spacing. Müller (1963) drew boundaries at the 1, 10 and 100 cm levels. Being fairly simple, this classification is in line with the adopted principle of equality of classification intervals on a logarithmic scale and can be recommended for use, say, in engineering geological studies of fracture systems.

As an alternative, boundaries to classes of discontinuity spacings with intervals also on a logarithmic scale (Table 1.4) have been proposed for national (Anon., 1970a, 1972, 1977, 1981a) and international (Anon., 1981b) adoption.

Block area distributions for two discontinuity sets

The ideas of discontinuity frequency and distribution spacing values along a scanline can be extended to block area frequency and the distribution of block area values in a plane (Hudson and Priest, 1979).

Table 1.4 Classification of fracturing based on fracture spacing

Fracture spacing (cm)	After Kriger and Preobrazhensky (1953)	After Belousov (1962)	After Müller (1963) recommended	After Deere (1968)	Fracturing After Rats and Chernyshev (1970)	After Jaeger (1973)	After Zhilenkov (1975) and USSR constr. norms and rules 11–16–76	USSR constr. norms and rules 11–44–78	After Anon., 1972, 1980. Brown, 1981
300		Very sparse		Very wide		Very sparse			
200						Sparse			Very widely spaced
100	I	Sparse	Sparse	Wide	Very sparse		Weak	Weak	Widely spaced
60		Medium thick				Medium			
50									
30	II	Thick	Medium thick	Moderately close	Sparse		Medium	Medium	Moderately widely spaced
20		Very thick							
10			Thick	Close	Dense	Thick			
5	III	Close		Very close	Very dense	Very thick	Intense	Intense	Closely spaced
3			Thick						Very closely spaced
2	IV		Very thick				Fairly intense		
1									Extremely closely spaced

In a rock mass containing discontinuity systems, if a given number, n, of discontinuities intersect a unit area of a scanplane, there is a minimum and maximum number of areas that can be bounded by the discontinuities. If the discontinuities are parallel, with no intersections, the minimum number of areas, $n + 1$, is produced. If, on the other hand, all the discontinuities mutually intersect, the maximum number of areas, $1 + [n(n + 1)/2]$, is produced. If n is 20, the minimum number of areas is 21 and the maximum is 211.

If a rock mass is divided into discrete blocks, a discontinuity value will represent a block edge length if the scanplane is parallel to one of the discontinuity sets. With two orthogonal discontinuity sets with negative exponential spacings, a tile pattern of regular areas is produced. If on the other hand the two sets are not orthogonal, the tile pattern will be parallelograms. The distribution of areas is the same in both cases, and therefore it is sufficient to analyse the area distribution for two orthogonal sets to provide the area distribution for two discontinuity sets inclined at any angle φ. Hudson and Priest (1979) show how the probability, $P(A \leq a)$, that an area A will be less than a given area a, for negative exponential spacing values, is:

$$P(A \leq a) = \int_0^\infty \lambda_1 e^{-\lambda_1 x} \left(1 - e^{-\lambda_2 a/x}\right) dx$$

where λ_1 and λ_2 are the discontinuity frequencies along the x and y axes respectively.

Apparently, the distribution of areas is not sensitive to assumptions on discontinuity orientations, and the random spacing is similar for random orientation of discontinuities and orthogonal discontinuity sets.

Cumulative block area proportion curves have been produced for two orthogonal discontinuity sets and for a totally random discontinuity geometry. The vertical axis of the graph (Figure 1.7(a)) is the percentage area of a rock surface that consists of areas less than a specified area given by the value on the horizontal axis.

The applicability of the block area distributions depends on the proportion of persistent discontinuities. As has been discussed earlier, some discontinuities may not extend completely across a rock mass, but terminate in intact rock or at another discontinuity (Figure 1.6(b)). If all discontinuities terminate at other discontinuities, the rock mass is divided into discrete blocks and the distributions will apply.

Block volume distributions for three discontinuity sets

Three mutually orthogonal discontinuity sets will produce a box pattern in three dimensions. The probability density distribution of volumes can be determined by a direct extension of the method described for the determination of the distribution of areas. This can be done for the simple case of uniform distribution of edge length spacings, but the block volume distributions produced by negative exponential distributed edge length spacings present, according to Hudson and Priest (1979), an intractable problem and have been determined by numerical simulation.

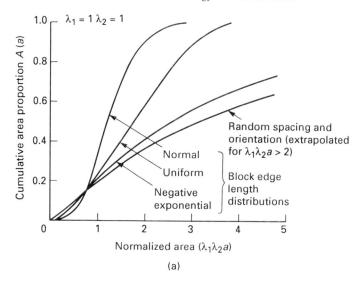

(a)

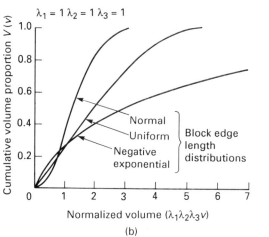

(b)

Figure 1.7 (a) Cumulative block area proportion curves for two orthogonal discontinuity sets and for a totally random discontinuity geometry (from Hudson and Priest, 1979, Figure 13). (b) Cumulative block volume proportion curves for three orthogonal discontinuity sets (from Hudson and Priest, 1979, Figure 16)

Cumulative volume curves have been produced from which the percentage volume of the rock mass that consists of block volumes less than a specified value can be determined (Figure 1.7(b)). From the curves a value of three-dimensional RQD can be established for normal, uniform, as well as negative exponential block edge length distributions.

Indirect determination of degree of fracturing

Degree of fracturing can be indirectly determined by specific water absorption, elastic wave velocity of rock material and the rock mass, and so on (see Part II for fracture investigation techniques).

Volume of voids in a rock mass

The porosity of rock materials in the upper few hundred metres of the Earth's crust rarely exceeds 10%, making up less than 1% on average (Rats and Chernysev, 1970). According to Müller (1963, p.95) porosity in rock materials rarely exceeds several per cent and is generally a few tenths of a per cent.

In a rock mass, the open fractures present contribute significantly to the volume of voids. If both average block size and the average aperture of open fractures are known, it is a simple matter to estimate the ratio of the volume of voids to the total volume of the rock mass, in other words the porosity. The porosity of rock materials is disregarded, although it may be significant in some rock types, for example in lightly cemented coarse-grained sandstones.

The practical relevance, in rock engineering, of rock mass porosity values due to systems of uniformly open fractures is likely to be of low importance as the geohydrology of the mass is dominated by a few fractures with high apertures (see Chapter 10).

If required, the range of variation of porosity values may be classified on a logarithmic scale as follows:

Descriptive term	Porosity (%)
Very low porosity	< 0.01
Low porosity	0.01–0.1
Medium porosity	0.1–1.0
High porosity	1.0–10
Extremely high porosity	> 10

1.5 Fracture filling

In terms of filling with mineral matter, fractures are divided into open, filled (with some loose material), and healed (by vein formation). The latter are not important in engineering geology and hence are not studied, whereas in mining geology they are the major target for exploration. Of great practical value in engineering geology are studies in permafrost areas of ice veins in fractures (Figure 1.8(a)), which greatly affect the strength of a rock mass on thawing.

The proposed classification of fillers (Table 1.5) is based on well-known concepts reflected in the Russian engineering geological literature. Skvortsov and Fromm (1970) divided fractures into those filled with the products of disintegration of the host rocks and those filled with the products of deposition from solutions. By classifying fillers primarily on their mode of origin, the following general types can be identified:

Figure 1.8 Fracture filler, granites, Kolyma Hydro foundation Adit 777, depth 40 m: (a) fracture with veined ice; (b) filler is coarse-grained sand

- tectonic breccias;
- mineral deposits caused by hydrothermal solutions;
- weathering products; and
- inwashed sediments.

It should be noted that there are no fillers of technogenic origin, if introduced grouts and other sealants are excluded, and also that the third and fourth classes tend to overlap.

The classification chart in Table 1.5 was drawn up on the general principle of lithology that rocks are divided on mode of deposition into: mechanical or detrital, chemical, and organic. Organic fillers are fairly rare in fractures yet they are included in the classification to provide a deeper insight into fracture filling.

Table 1.5 Classification of fracture fillers by origin

Deposition of fracture filler	Description of filler based on material	Composition and properties of fracture filler
Chemical or physico-chemical	Magmatic	Rock healing fracture solidly
	Hydrothermal and pneumatolytic	Rock healing fracture
	Hypergene	Colloidal formations which cause fracture narrowing or healing
	Artificial	Chemical grout infilling fracture
Mechanical	Tectonic	Mylonite, fault breccia. Compact, impervious, low-strength, slightly compressed
	Hypergene	Clastic or clay, loose rocks. Impervious, low-strength, compressed
	Artificial	Cement grout infilling fracture
Organic	Phytogenic	Plant roots, rotting residues. Permeable medium, facilitates weathering
	Zoogenic	Organic residues and rotting products washed into fractures. Weakens rock mass and facilitates weathering

Effect of filling on physical behaviour

Filled, and partly open fractures, vary widely in their physical behaviour, particularly in terms of their shear strength, deformability and permeability. Both short-term and long-term behaviour, which may be quite different, depend on many factors, including the following:

• mineralogy of filling material;
• grading or particle size;
• over-consolidation ratio;
• water content and permeability;
• previous shear displacement;
• wall roughness;
• width; and
• fracturing or crushing of wall rock.

Weathering grades

Filled discontinuities that have formed as a result of differential weathering along pre-existing discontinuities may have fillings composed of decomposed rock, or disintegrated rock. In the West, weathering classifications developed from the work of Ruxton and Berry (1957) have been recommended (Dearman, 1974, 1976; Anon., 1981a), and the relevant type of weathered rock material should be recorded:

- Decomposed: the rock is weathered to the condition of a soil in which the original material fabric is still intact, but some or all of the mineral grains are decomposed.
- Disintegrated: the rock is weathered to the condition of a soil, in which the original material fabric is still intact. The rock is friable, but the mineral grains are not decomposed.

In field studies, the condition of the wall rock including the filling and associated fractures is particularly relevant.

Characteristics of fillers

Fillers of organic origin differ greatly in composition. Organic matter breaks down in fractures to form acids promoting weathering. Chemical deposits in fractures are represented by calcite, vein quartz, amorphous varieties of silica, sulphides, sulphates. A fracture is generally well healed by chemical deposits. It has a minor role to play in mechanical weakening of a rock mass because of the cohesion between its walls. However, this cohesion due to the presence, for example, of calcite, gypsum and limonite does not exceed cohesion in the rock. That is why a rock mass is often disintegrated along the fractures healed by chemical and physico-chemical deposits. Sometimes, chemical and physico-chemical deposits accumulate only as a film on the surface of the fracture wall.

Figure 1.9 Gaping fissures in granodiorite originating from suffosion of weathered tectonite (Erdenet Mining and Concentration Complex, Mongolia), photograph by M. I. Pogrebisky

Fillers of mechanical deposits are sand, silt and clay particles which accumulate in fractures carried along by water or having been formed in situ. An in situ filler is more dense than the one carried along with water. For example, tight coarse gruss sand in granite results from weathering of cataclastic rock. Cataclasis and weathering affect mostly lenticular blocks of rock which occur between contiguous fractures. As a result, sand lenses (Figure 1.8(b)) confined by strong rock, appear in granite. With a sand filler subjected to suffosion, gaping joint-fissures develop in a rock mass, the opening being maintained by individual strong rock blocks (Figure 1.9). Fillers of mechanical origin formed by weathering have low density, less than that of mylonites and Quaternary surficial deposits. Large fragments and fracture wall roughness inhibit compactibility of mechanical sediments. Fillers of mechanical deposition often accumulate in some narrowed portions of fractures leaving their widened portions open.

Chapter 2
The main genetic types of rock fractures

2.1 Genetic classification

Little has changed since M. V. Rats wrote: ' . . . so far it is impossible to provide an adequate genetic classification of fractures' (Rats and Chernyshev, 1970). That is why it is not possible to distinguish, in many cases, between lithogenetic and tectonic fractures, weathering and release fractures, lithogenetic and weathering fractures, planetary and fold fractures, and so on. The problem is so complicated that less interest has been shown lately in genetic classifications, although they are fairly important for the following reasons, namely: (i) the genetic terminology reflects the experience in studying fractures which has been accumulated by several generations of geologists; (ii) genetic classifications are essential in forecasting fracturing in regional studies based on investigations of structural geological conditions; (iii) they are required to extrapolate and interpolate observational data on fractures between investigation boreholes in engineering geological studies and geological exploration.

Genetic classifications present many problems since it is considered that the role of inheritance in fracture evolution has been underestimated. Up to now genetic types of fractures have been treated individually. When studying fractures of any genetic type, an example was selected which had almost no fracturing changes brought about by processes unrelated to the development of that particular type of fracture. If such changes did take place, they were disregarded as a handicap. However, in practice this is impossible, say when studying a rock mass from the standpoint of engineering geology. It has to be admitted that most fractures are characterized by inherited evolution. That is why they generally bear traits acquired in lithogenesis, tectonic deformation, weathering and sometimes construction activity. To which genetic type does a fracture belong? In practice, it is usually attributed to the type which manifests itself most strongly as the geologist sees it. Hence some problems arise. It is at this point that a genetic classification of fractures is very essential in the sense that it enables their specific features to be disregarded.

Analysing the genesis of fractures in a rock mass, it is advisable to trace the evolution of the rock mass and to single out stages of fracture formation or transformation. These stages generally include (i) primary fracturing in lithogenesis; (ii) tectonic deformations; (iii) weathering; (iv) construction activity (technogenesis). Each stage, except naturally the first, is characterized by inherited evolution of earlier fractures. New fractures may also develop. Three types of fractures can be singled out, namely (i) mostly healed fractures preserved from an earlier stage; (ii) inherited old fractures transformed at a new stage; (iii) newly formed fractures. The latter include fractures which bear no traits inherent in fractures from the earlier stages of the geological history of the rock mass. In some cases newly formed fractures may include highly transformed earlier fractures. Indeed, there can be no newly formed fractures unless earlier microfissures merge. At this point a parallel can be drawn with metamorphism: a newly formed rock inherits certain features from an original rock, but properties of the latter disappear.

All genetic classifications have been made only for newly formed fractures. That is why when studying fractures in a particular rock mass, it is necessary, first and foremost, to single out newly formed fractures for each stage and then consider transformation of fractures at subsequent stages. A case in point is given at the end of the section.

Consider now the traditional genetic classification of fractures bearing in mind that it covers only newly formed fractures (Table 2.1).

An analysis of known genetic classifications of fractures (including Cloos, 1923; Müller, 1963; Balk, 1937; Sander, 1948; Belousov 1962; Kolomensky, 1952; Zolotarev, 1962; Rats and Chernyshev, 1970; and many others) has shown that they are mostly based on fundamental geological evidence which has never been mentioned. This evidence includes the energy responsible for fracturing, time, and place of origin. A distinction is, therefore, made primarily as to the origin and direction of energy flow responsible for fracturing. Thus, endokinetic fractures can be singled out which develop on account of energy stored in a rock, and exokinetic fractures which result from external effects. Another distinction is based on the relationship of fractures to a particular stage of rock

Table 2.1 Genetic classification of fractures

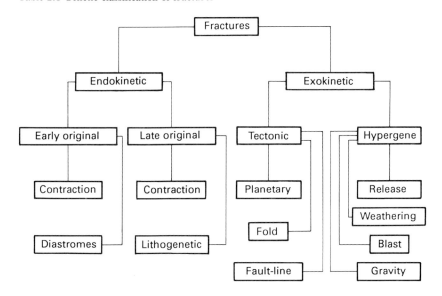

evolution, namely geological time. Four types of fractures are involved in this aspect, namely (i) original fractures which develop in mobile rock material; (ii) original fractures which develop in a sediment or a melt once the flow has stopped; (iii) tectonic fractures which develop in a rock following lithification (crystallization) under the effect of crustal movement; (iv) hypergene[*] fractures which develop in a rock at the final stages after it has been exposed at the surface in the zone of weathering, or by the release of elastic stresses.

Each of these four types of fracture is subdivided as to its place of origin determined by a particular lithological and structural environment. Among original fractures there are those in abyssal, hypabyssal and effusive rocks, as well as in sedimentary and pyroclastic rocks. Tectonic fractures include those in horizontal or gently inclined stratified formations (planetary), and fold and fault-line fractures. Among hypergene fractures there are release fractures and weathering joints. Also treated are specific fractures resulting from blasting or impact.

Further distinction can be drawn as to particular forms of geological structures. Several dozens of subtypes can be singled out, for example fault-line fractures may be subdivided into thrust and fault fissures, while contraction fractures in hypabyssal bodies may be subdivided into dyke, sill, laccolith, lopolith, and apophysis fractures. The above fractures have their specific features depending upon configuration, spatial position and kinematics of formation of faults or intrusive bodies. Finally, types of fractures are classified on the basis of rock composition. There are many classes which cannot be fitted into a single system. For example, weathering joints in granites, basalts, limestones, schists, marls, sandstones – each feature a specific morphology. These will be treated further later in this book.

A genetic code for fractures

Consider the genetic types of fractures in the Bolsheporozhsky granite batholith on the Kolyma river. The rock mass has undergone original contraction fracturing, tectonic deformations during orogenesis, weathering and construction deformations. At each stage newly formed fractures developed and earlier fractures experienced inherited evolution. Formation of the fracture system is tabulated in Table 2.2. Types of fractures at different stages in the geological history of the rock mass are shown. Only contraction fractures (C) are listed at the petrogenetic stage; at the orogenic stage these come together with tectonic fractures (T) and contraction (CT) transformed in tectonic movements.

There are four classes and 15 genetic types of fractures in the rock mass at the present stage, for example 1 – original unaltered fractures; 6 – original fractures transformed as a result of tectonic and construction processes; 15 – fractures newly formed in construction. Not all of them can

---

[*] Hypergene; synonym supergene. Suggesting an origin literally 'from above'. It is used almost exclusively for processes involving water, with or without dissolved substance, percolating down from the surface. The chief of these is chemical weathering involving decomposition and solution.

Table 2.2 Genetic code of fractures

Stages in rock mass geological history	Fractures in a rock mass														
	1	2	3	4	5	6	7	8	9	10	11	12	13	14	15
Anthropogenesis	C—	C—A	C-H-	C-HA	CT—	CT-A	CTH-	CTHA	-T—	-T-A	-TH-	-THA	-H-	-HA	—C
Hypergenesis	C—		C-H		CT-		CTH		-T-		-TH		-H		
Tectogenesis			C-				CT				-T				
Petrogenesis					C										

Notes:
C: contraction (original) fractures.
T: tectonic fractures and fractures transformed in tectogenesis.
H: hypergene fractures and fractures transformed in hypergenesis.
A: anthropogenic blast fractures and fractures transformed in the process of construction

be found in foundation excavations. So, tectonic fractures not transformed in hypergenesis were not observed since weathering took place in all areas along fracture zones. Fractures of types 9 and 10 can be encountered at depth, whereas fractures of types 7 and 8 are most common since they feature inherited evolution. These include original fractures transformed as a result of tectonic and hypergene processes (7) and the same (8) transformed as a result of tectonic and hypergene processes, and also in construction. Since fractures also undergo inherited evolution at the technogenic stage, types 7 and 8 are to be considered as those most important from the practical point of view. For example, 1 – original fractures not transformed in tectonic processes, weathering and construction are mostly short veins healed by hydrothermal minerals; 2 – same fractures with no weathering traces but open as a result of blasting or release of elastic stresses.

The digital designation of a genetic type is short but difficult to perceive. The verbal definition is too long: it takes 14 Russian and 10 English words to define fractures of type 1. It is advisable to denote the genetic type of a fracture by a formula with so many letters as the number of fracturing stages it went through, stages in which the fracture remained unaffected being marked by a dash. For example, C – – A is for genetic type 2, CTHA is for genetic type 8, – – H A is for genetic type 14.

Information on fracture alteration at a particular stage may be incorporated in the formula if required. To do this, fractures are ranked according to the degree to which they have been altered. The rank number can be indicated as a digital index of a letter standing for a particular stage. A method of fracture transformation can also be shown. For instance, CT_1wH_2- is to be read as: original fracture slightly transformed by tectonic deformations, substantially transformed by weathering and not affected by construction processes. The proposed code facilitates storage of genetic data in information retrieval systems.

2.2   Original fractures

Original fractures are formed in a crystallized melt or a lithified sediment on account of energy stored in intrusion, effusion and sedimentation. The nature of original fracture is, therefore, governed by rock composition, mode of occurrence and rate of petrogenesis. Original fractures may also be affected by the stressed state and dynamic setting of neighbouring rock masses. Fractures are formed in intrusive rocks solely in the presence of country rocks. Original fractures develop most commonly in a horizontal lava sheet or sedimentary layer. With no effect produced by country rocks, stresses are due to contraction and gravity forces. Contraction in isotropic material is responsible for a spherical stress tensor, and gravity force for a vertical principal compression stress. With the stress fields combined, a stress ellipsoid acquires the form of an ellipsoid of revolution with a short vertical axis and equal axes in the horizontal plane. The ellipsoid has a circular cross-section. Under these conditions with combined tension and increasing contraction stresses vertical tension fractures mostly develop. Azimuthally, these are arranged arbitrarily on account of equal principal normal stress in the horizontal plane. Structural heterogeneities in a sediment or a melt and temperature field fluctuations affect the equality of tensile stresses in the horizontal plane. Vertical fractures are, therefore, arranged differently giving rise to a system of non-parallel slits closing downwards (Figure 2.1). Once vertical fractures are formed, rock stresses rearrange, the vertically directed principal normal stress being at a minimum.

Sometimes with rock blocks squeezed in vertical fractures, contraction stresses exceed gravity stresses, thus giving rise to horizontal fractures which run parallel to each other and form a set. It is unlikely that contraction stresses exceed vertical pressure with depth; therefore the horizontal fracture spacing in lava sheets and hypabyssal sills increases with depth. In sedimentary rocks, horizontal fractures generally coincide with layer interfaces, their spacing being adjusted in the process of sedimentation.

Fracture width and the breadth of polygons depend upon the amount of energy stored in a sediment or a melt. For example, the more bound-water retained by clayey particles in a sediment, the more energy is liberated on dehydration and the higher is rock mass porosity. The same relationship holds for melt temperature and porosity. That is why porosity due to original fractures is constant in one and the same facies (Tolokonnikov, 1966). The vertical fracture spacing in thick lava sheets decreases with depth (Levinson-Lessing, 1940; Ter-Stepanian and Arakelyan, 1975; Aydin and DeGraff, 1988). The stress field under consideration enables this feature of the original fracture system to be explained by the fact that vertical compression stresses increase with depth.

There is ample geological and other literature on the shape of polygons and junction of vertical fractures. Basalts are known to display mostly pentangular or hexangular jointing (Tolokonnikov, 1966). Almost regular orthogonal sets with quadrangular jointing can be observed occasionally. Studying systems of frost clefts, Dostovalov (1959) came to the conclusion that orthogonal fractures are typical of homogeneous rock masses, whereas hexahedrons are formed in heterogeneous rock masses in the event of fracture curvature. When describing original fracturing, Rats and Chernyshev (1970) proceeded from the assumption that fractures originated simultaneously at a great many points throughout the rock mass, and also grew and closed.

Figure 2.1(a) shows an incomplete system of original fractures. Independently developing sections can be clearly seen. The development of original fractures throughout the rock mass at one time gives rise to hexahedral joint columns (Smally, 1966). However, in actual rock masses the average number of faces of joint columns is usually less than six (Table 2.3). Smally attributes this to the fact that very short sides of polyhedrons existing on models are actually not observed under natural conditions.

Original fracture systems are also peculiar in the way fractures intersect each other. The fracture sets under consideration are characterized by triple intersections (see Figure 2.1 and Table 2.4), which is in line with the concept of independent evolution of individual portions of the set. Indeed, if a source of the stress field is located within a polygon the generated fracture will not extend beyond the one already formed, i.e. the polygon

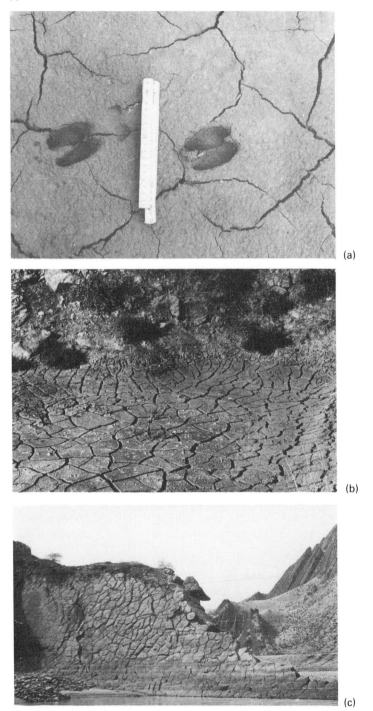

(a)

(b)

(c)

Figure 2.1 Original polygonal fracture systems in sand and clay sedimentary deposits in Tajikistan (Nurek Hydro reservoir), photograph by M. I. Pogrebisky: (a) interrupted set of desiccation fractures on the surface of solidified mudflow; (b) continuous set of desiccation fractures in clay sediment; (c) original fractures in Cretaceous sandstone

Table 2.3 Data on the number of polygon faces in original fracture systems (Rats and Chernyshev, 1970)

No.	Location	3	4	5	6	7	8	Average number of faces	Observed number of faces
		Percentage of polygons with the number of faces							
1	Pleistocene basalt flow in the moon crater area, Idaho	–	28	56	16	–	–	4.9	50
2	Pleistocene basalt flow Dancemoore, California	0.5	14.3	46.0	33.5	4.5	1.0	5.3	200
3	Basalts 'Roads of Giants', Northern Ireland	–	4.5	35.0	51.0	9.25	0.25	5.7	400
4	Basalts Devils Postnile, California	0.5	9.5	37.5	44.5	8.0	–	5.0	400
5	Basalts, Sidenham, Australia	–	3	22	40	22	13	6.2	
6	Basalts in Armenia	1	13	36	36	13	1	5.5	
7	Andesite lavas in Georgia	–	4.4	32	63.6	–	–	5.6	
8	Sediment in the Vakhsh river valley	10	27	17	27	12	7	5.0	150
9	Same	10	27	35	19	7	2	5.0	200

boundary. With an external source for the stress field, fractures evolve differently. The external force gives rise to a relatively homogeneous stress field in all rock mass blocks, and fractures are traced from one block to another. Therefore, tetraradiate intersections (Table 2.4) with tetragonal 'cells' can often be observed in the plane section of tectonic fracture sets. The same results are obtained with external stress fields superimposed on contraction stress fields. That is why contraction fractures in batholiths are closer, in terms of geometry, to tectonic fractures than to original fractures described above.

According to Cloos (1923), granite batholiths display transverse and longitudinal, steep and flat, as well as diagonal joints with respect to flow texture. They form four major systems of original fractures and are denoted by *Q*, *S*, *L*, and *D*. Each system has some peculiar features connected with its arrangement relative to the axes of principal normal stresses which run in a particular way with respect to flow texture. Cloos' ideas on the subject have met with both support (Pek, 1939; Balk, 1937; Kriger, 1951) and criticism (Yeliseyev, 1953).

Engineering geological studies of fractures in granite masses (Kolyma, Mongolia, Central Asia) provide convincing evidence that Cloos' systematics reflect the principal features of original granite fracturing. Deviations from the scheme are related to the anomalous or unstable effect of country rocks. On the whole, as the contact is approached deviations from the Cloos scheme become ever more apparent. Generally, dispersion of all fracture parameters tends to increase as does heterogeneity, and hence deviations from any scheme become very apparent.

The original fracturing of hypabyssal bodies has much in common with that of effusive rock masses and plutonic intrusions. This may be easily

Table 2.4 Fracture intersection

Number of rays in intersection	Percentage of junctions with a specified number of rays									
	Original fractures								Tectonic fractures	Hypergene fractures
	lava sheet			granite batholith	desiccation in silt, four areas				fold fractures in siltstones, Nurek (Central Asia)	blast fractures in granites (Kolyma)
	glass	$l \approx 10$ m	$l \approx 1$ m		1	2	3	4		
1	0.6	0.8	2.7	6.7	1.5	3.3	4.2	6.5	10.	1.4
2	0	0	0	2.9	0	0	0	0	0	12.8
3	77.5	71.8	72.0	60.0	83.2	88.5	86	75	18	72.8
4	21.5	23.9	12.0	32.6	14.8	8.2	8.4	18.5	72	11.4
5	0.4	2.3	1.8	0.4	0	0	0	0	0	1.4
6	0	0.8	0.6	0.4	0.5	0	0	0	0	0
7	0	0.4	0.6	0	0	0	0	0	0	0
8	0	0	0.3	0	0	0	0	0	0	0

understood since the hypabyssal facies is intermediate between plutonic bodies and effusives as regards crystallization and fracturing conditions. Fracturing of hypabyssal and plutonic intrusions has the following common features: (i) development, as a rule, of three systems of orthogonal fractures; (ii) hydrothermal fracture filling; (iii) near-contact variations of fracturing. The same systems of original fractures in hypabyssal bodies are related to effusives by means of pronounced columnar jointing and locally by spheroidal and irregular blocky jointing.

Original fractures in dykes are mostly perpendicular to the contact and form a polygonal system whose configuration is similar to that in sedimentary rocks. They look like fractures in sandstone (Figure 2.1(c)). It is interesting to emphasize morphological similarity of fracture systems reflecting similarity between the stressed state of sedimentary and igneous rocks during fracturing. Of course, there is a difference between fracture systems in sedimentary and igneous rocks, which stems from the general rock mass structure. A sedimentary rock mass is stratified, and original fractures are localized in individual layers at right-angles thereto, their length being proportional to layer thickness. A single fracture usually intersects one to three layers at most. The original fracture spacing in sedimentary rocks and fracture width are related to layer thickness. The average width of fractures in a silt layer 1.1 m thick is 7 cm reaching a maximum of 12 cm, whereas in a silt layer 0.2 m thick fracture width is about 1 cm and in a silt layer about 0.01 m thick the width of desiccation fissures is some 0.1 cm.

In igneous as opposed to sedimentary rocks, the structural patterns of original fracture systems are more varied, which stems from a variety of body forms in igneous rocks that might be similar to a layer, in which case original fractures of igneous rock bodies are similar to those in sedimentary rocks. The configuration of igneous rock bodies ranges from the simple pillow lava of an underwater flow with spheroidal jointing (Figure 1.3(a)) to intricate interlacing of individual lava jets where fractures follow no pattern. In intrusive bodies and thick effusives, fractures at the surface of the igneous rock body are generally perpendicular to the contact or flow surface. As they run deeper into the body they tend to assume a vertical position. Thus curved columnar jointing and fan-shaped arrangement of fractures can be observed at the bent contacts of igneous rock bodies.

2.3   Tectonic fractures

Tectonic fractures originate under the effect of external forces connected with relative displacement of masses in the Earth's crust. It is not the origin or magnitude of these forces that matters in studying the geometry of fracture sets. Of primary importance, tectonic forces are external and contraction forces internal with respect to a rock mass. This factor accounts for a basic characteristic feature of tectonic fractures, namely their arrangement in sets (Figure 2.2).

One fracture set may arise from original fracturing, but two or more sets always owe their origin to the external field of tectonic, mostly compressional, stresses. Several fracture sets in the original system may develop

(a)

(b)

Figure 2.2 (a) Regular system of rectilinear tectonic fractures and spheroidal system of weathering fractures on the surface of siltstone stratum, Baipaz Hydro, Vakhsh valley. Tectonic fracture spacing is about 30 cm. (b) Tectonic fractures on Kuli–Meyer fold limb, Dagestan

only if the contraction stress field and external tectonic field are combined during original fracturing. Several sets in the hypergenesis zone can often be observed during revival of the multi-set system of tectonic fractures. This most common feature of tectonic fractures is not always mentioned when describing fracturing.

Rectilinear arrangement

Another common feature of tectonic fractures, their rectilinear arrangement, is more typical of macro- and microfissures than of faults. Emphasizing these features of tectonic fractures, Kosygin (1974) wrote: 'Tectonic fractures are mostly rectilinear and follow a regular pattern.' A regular rectilinear arrangement of fractures is due to tectonic stresses constant in space and time. When intersecting geological boundaries are related to dramatic changes in rock properties, the regular rectilinear arrangement of fractures becomes affected. In homogeneous rock masses, tectonic fractures retain their rectilinearity for hundreds of kilometres.

For instance rectilinear tectonic faults several kilometres long can be seen in homogeneous Barremian limestones in Dagestan. A photograph (Figure 2.2(b)) taken from the top of a mountain shows a limb of a large anticlinal fold composed of a thick layer of massive limestone. Rectilinear parallel ravines, coinciding with tension fractures which cross the fold, run downhill. Shear fractures intersect these ravines at an angle of about 45°. The visible length of the fracture nearest the onlooker is about 1.5 km.

While emphasizing rectilinearity of tectonic fractures as their common feature, it should also be noted that ring faults, curved and split faults develop under certain structural geological conditions. They are related either to localization of stresses from intruding magma or a salt plug, or to the mechanical heterogeneity of a rock mass.

Each class of tectonic fracture has distinguishing characteristics studied in detail over the last decades. Consider planetary, fold and fault-line fractures in the sequence corresponding to ever greater localization of fracture systems and somehow reflecting their order of formation.

Planetary or general fractures

Planetary or general (Belousov, 1962) fractures are fairly widespread, forming, along with lithogenetic fractures, the background to fracturing in sedimentary rocks. They have the following features:

1. Two mutually perpendicular sets are formed at each point of a rock mass.
2. Sets of general fractures are traced along the section of platform deposits from the Precambrian to Quaternary (these have been observed in Quaternary travertines in the Caucasus and loams in Mongolia). Their arrangement is fixed and their spacing is determined by the composition and thickness of layers at a particular depth.
3. Fractures are more pronounced in strong rocks wherein they are mostly perpendicular to bedding and do not run beyond the boundaries of a single layer.

Figure 2.3 Planetary fractures in horizontally stratified formation of marl and siltstone; outcrop 12 m high

4. As layer thickness increases, so does fracture spacing. The last two features of planetary fractures may be clearly seen in Figure 2.3.

As opposed to lithogenetic fractures, planetary fractures feature a systematic arrangement which is not, however, always pronounced due to a large spread of fracture azimuthal values. They are also more rectilinear and have a greater vertical extent, i.e. possess all the features peculiar to tectonic fractures. A relationship between fracture spacing and layer thickness common to lithogenetic and general fractures was apparently noted for the first time by Bogdanov in 1947 and studied quantitatively by Rats in 1962 (Rats and Chernyshev, 1970), who showed that as layer thickness increases so does fracture spacing in accordance with the function:

$$a = rM^k$$

where: M is layer thickness;
r and k are empirical coefficients.

Fold fractures

Fold fractures originate in rocks apparently under the effect of the same stresses responsible for the folds, these stresses being transformed at each point of a rock mass. Based on many studies, the following conclusions can be drawn on fold fracture patterns. Fold fractures are grouped into sets. There are five principal sets arranged in a certain pattern with respect to the layer plane and fold axis, namely: set I (parallel to the layer plane), set II (perpendicular to the layer and fold axis), set IV (perpendicular to the layer and parallel to the fold axis), sets III and V (perpendicular to the layer and running at an angle to the fold axis). The angle between planes of fracture sets III and V is close to 90°, usually deviating from the above value by 10–20° (Figures 2.4–2.6); in many cases these sets include 70 to over 90% of fractures (Belousov, 1962). The above-mentioned sets are not always encountered in combination with each other. The difference between fractures in longitudinal compressed folds and transverse flexure folds is that some particular sets predominate in the angles between fractures and bedding plane. Normally, this difference is difficult to discern with fracture parameters varying over wide limits, and hence may be neglected in engineering geology.

Thus, the arrangement of fracture sets in a fold limb is related to the position of the stress axis during folding rather than to layer strike and dip. The location of these axes, or Sander coordinates, can be reconstructed on the basis of layer dip and strike and fold axis direction. For longitudinal compressed folds the axis of maximum principal normal stresses runs along the layer at right-angles to the fold axis, and the axis of intermediate principal normal stresses lies in the layer plane parallel to the fold axis. Fractures originating in the stress field are perpendicular to the layer and parallel or perpendicular to the fold axis rather than to the bedding strike or dip line, as noted in the literature (Kirillova, 1949).

Oblique fractures are also positioned in a particular manner with respect to the bedding plane and fold axis rather than to bedding strike and dip. When the fold axis is horizontal, as is generally the case in areas of open folds, the direction of principal normal stresses coincides with strike and dip directions. In this specific case Kirillova's version holds true. The generalization thus carried out has made it possible to devise a method of revolution which has found wide use in structural and engineering geology (see Chapter 8). The uncorrected method of revolution leads to erroneous conclusions that the fracture set pattern lacks unity in different portions of the fold. Gulamov (1975) arrived at the same conclusion, calling for the need to adjust the method of revolution.

There are primary and secondary fracture sets which include most fractures and individual local fractures respectively. Even the primary sets are not always found in combination. Some folds display only three primary fracture sets: bedding fractures and some two out of the four primary fracture sets, two tension fracture sets running parallel and perpendicular to the fold axis or two shear fracture sets perpendicular to the bedding and forming an angle close to 45° and 135° with the fold axis in longitudinal compressed folds.

The arrangement of sets in paired systems has been ascertained, particularly in folds in the Kerman coal deposit, Iran (Table 2.5). The dense arrangement of fractures in a set perpendicular to the fold axis (II) is spatially associated with a dense arrangement of fractures in another tension fracture set parallel to the fold axis (IV). The correlation coefficient (0.66) reflecting this relationship is of positive sign. Fracture spacing in a pair of conjugate shear fracture sets (III and IV) also varies in direct proportion to each other. By contract, in cases where tension fractures become denser, shear fractures are more sparse and vice versa. This manifests itself as negative values of correlation coefficients between shear and tension fracture densities, e.g. for sets II and V or IV and V. For

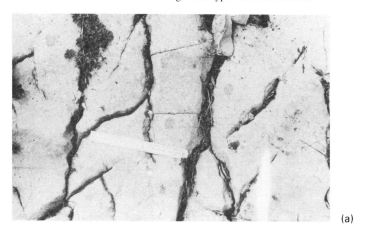

(a)

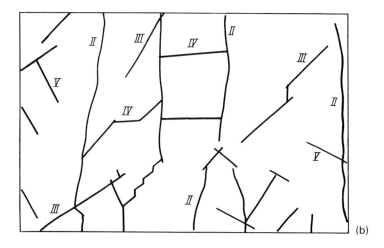

(b)

Figure 2.4 Four systems of tectonic fractures perpendicular to bedding on the limb of the Kuli–Meyer anticline: (a) photograph of the set of fractures enlarged by solution; the rule is 15 cm long; (b) line drawing of the occurrence; (c) circular diagram

constructing Table 2.5, use has been made of fracture density measurements taken on outcrops at a total of 33 stations, each one measuring about 200 fracture spacing, 200 dip azimuths and 200 dip angles. Average fracture spacings in sets II, III, IV and V have been calculated for each station. The correlation coefficient has been calculated from average fracture spacings in a set. For instance, all 33 values of average fracture

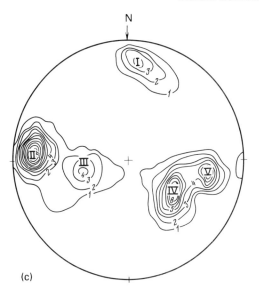

(c)

Figure 2.5 Fold limb relief related to tectonic fractures in Albian limestones in the Avarsky valley

spacings in set II were taken as y and those in set IV were taken as x. In this case the correlation coefficient was 0.66.

Fold fracture density has three features most essential to engineering geology, namely (i) fractures perpendicular or parallel to bedding strike are spaced more closely than oblique fractures; (ii) spacing between

(a)

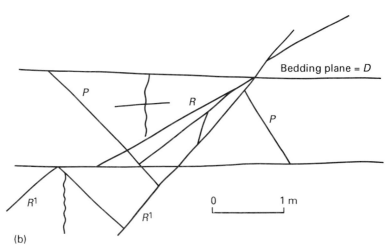

(b)

Figure 2.6 Tectonic shear fractures inclined to the bedding of Jurassic siltstones in the crest of an anticline, Avarsky Koisu valley: (a) photograph; (b) line drawing

Table 2.5 Correlation matrix of fracture density in different sets in fold

Set No.	II	III	IV	V
II	1.0	–	–	–
III	– 0.39	1.0	–	–
IV	0.66	– 0.24	1.0	–
V	– 0.48	0.74	– 0.61	1.0

fractures perpendicular to the layer depends on its thickness; (iii) out of four fracture sets perpendicular to the layer, two approximately orthogonal sets are most common.

Fault-line fractures

Fault-line fractures are encountered in the vicinity of tectonic faults (Budko, 1958; Mikhailov, 1956). As to the width of the feathering zone, in which the fractures appear to the fault as the barbs of a feather to its shaft, and variations in density in fault-line fracturing zones, these have been studied in less detail.

Fault-line fractures owe their origin to the stress field forming the fault, and the location of stress axes determines that of advance fractures and the fault which eventually evolves from them. All the faults are formed by merging shear fractures (Tetyayev, 1940). According to Mohr's theory, supported by geological observations (Tchalenko, 1970; Stoyanov, 1977; Pogrebisky and Chernysev, 1974), shear fractures give rise to two fracture systems at an angle of $\pi/2 - \varphi$ to each other, where φ is the angle of internal rock friction.

The so-called 'advance fractures' mentioned above are more commonly referred to as Riedel shears. Shear zones parallel to bedding were discovered in clay beds exposed in the excavations for the Jari dam on the Mangla Dam Project in Pakistan. The strata, alternating beds of compact clays and sandstones or boulder conglomerates folded into an asymmetrical anticline, have been affected by bedding plane slip induced by concentric folding, forming shear zones. Skempton (1966) has studied those tectonic shear zones both in field exposures and by experimental deformation of clay in which the tests were carried to large displacements.

Observations on faults, bedding plane slips and landslides lead to the conclusion that slip in shear zones is not always confined to a single surface or to conjugate sets of single surfaces. In shear zones, shearing has occurred along many slip surfaces distributed throughout a zone, varying in width from a few millimetres to over a metre in small geological structures. When fully developed, a shear zone has definite boundaries within which the style of shearing contrasts sharply with the fracture patterns in the adjacent rock, even though this may contain various accommodation fractures, thrust shear joints, and so on.

A schematic representation of a shear zone of the classic type (Figure 2.7a) involves displacement shears, Riedel shears, thrust shears and tension fractures. Shears are referred to the conventional system of rectangular axes in which a is the direction of movement and b lies in the plane of shear, with c at right-angles to this plane (Figure 2.7(a)).

Slip surfaces of one important set lie parallel or subparallel to the ab plane. Skempton defined them as 'displacement shears' and denoted them by the letter D.

The term displacement shears is limited to those particular slip surfaces which follow the direction of general displacement. If the movements are large slip becomes concentrated along one, or sometimes a pair, of these displacement shears, which can then be distinguished as 'principal displacement shears' or, 'principal slip surfaces'.

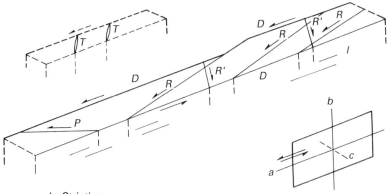

I = Striation
D = Displacement shear
R = Riedel shear
P = Thrust shear
T = Tension fracture

Reference axes

(a)

(i)

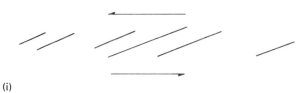

(ii)

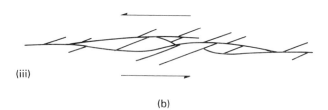

(iii)

(b)

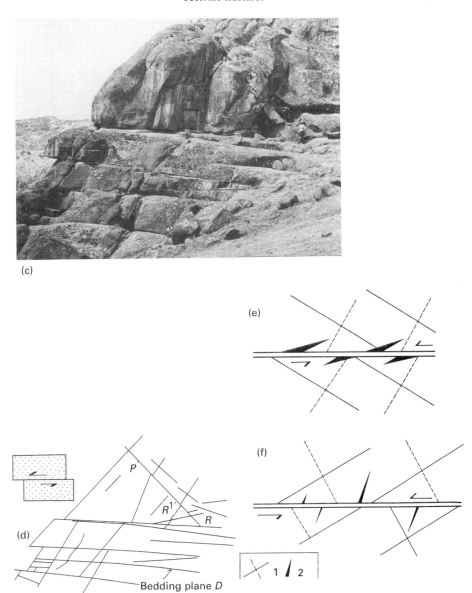

Figure 2.7 Feather joints and Riedel shears associated with faulting: (a) schematic representation of a shear zone of the classic type (after Skempton, 1966, Figure 1); (b) sketch showing three successive stages in the development of slip surfaces in clay subjected to simple shear (after Skempton, 1966, Figure 5); (c) feather joints associated with the Sarikamar thrust fault, Nurek reservoir, in massive Neogene sandstones. The displacement vector is in the plane of the photograph, and three systems of feather joints are perpendicular to the plane of the photograph. The outcrop is 15 m high (photograph by M. I. Pogrebisky); (d) line drawing of the fractures in the outcrop; (3) theoretical fracture pattern resulting from strike slip faulting with additional tension (after Gzovsky, 1975); (f) the same with additional compression: 1 – shear fractures, 2 – tension fractures. The area of outcrop is 30 × 40 cm

Riedel shears, slip surfaces of another set, typically lie *en echelon*, inclined at 10° to 30° to the *ab* plane, with the acute angle always pointing against the direction of relative movement, as indicated by R in Fig. 2.7a. Ideally they are accompanied by a conjugate set R^1.

In simple shear the Coulomb failure criterion predicts that the R surfaces are inclined at $\varphi/2$ to the *ab* plane, where φ is the angle of internal friction. In the general case where normal stresses are present as well the situation is more complicated. The conjugate set, however, should invariably make a dihedral angle of $(90 - \varphi)$ with the main R shears. After formation the slip planes rotate, with continued shearing, and the geometrical relationships are altered accordingly.

Thrust shears have an orientation opposite to the Riedel shears, in the position approximating to a mirror image. The acute angle with the *a* axis points in the direction of relative movement. Thrust shears are denoted by the letter P in Figure 2.7(a).

The combined effect of the various sets of slip surfaces is to divide the shear zone into numerous lenses, bounded by R and D shears.

Experiments on the development of slip surfaces in clay subjected to simple shear have revealed that five main stages are involved. The first stage is one of continuous, non-homogeneous strain. Riedel shears are formed in the second stage (Figure 2.7(b(i))), lying *en echelon* at an angle of $\varphi/2$ to the *a* axis. With continued displacement a third stage is soon reached at which further slip along the R shears is no longer kinematically possible, and new slip surfaces are developed parallel to the *a* axis (Figure 2.7(b(ii))). These are the displacement shears, D.

With greater movements the D shears extend and eventually, in the fourth stage, some of them link up to form a principal shear surface. This surface is undulating (Figure 2.7(b(iii))) since the D shears involved were not all in the same plane. A few thrust shears, P, are usually present in this stage.

In the fifth stage, not illustrated, with still greater displacements the principal slip surface undergoes appreciable flattening, accompanied by minor high-angle shears. Thus five sets of fault-line fractures are located so that the intersection lines are parallel to the major fault plane and perpendicular to the direction of relative displacement (Figure 2.7).

The large number of fault-line fracture sets and their great dispersion in each particular set leads to coalescence of the sets at long-lived faults. The circular diagram of fracturing shows feather joints and fractures merging together to form a continuous field of points referred to as a 'feathering belt' (Figure 2.8); some faults feature not only belt fractures, but also those perpendicular to the belt. In circular diagrams, their position coincides with the fault surface pole. Thus, the prominent system of fault-line fractures with a large number of feathering sets is geometrically similar to the system of original fractures originating irrespective of the external stress field. This similarity will be used in developing a single geometrical classification of fracture systems.

Feathering joints give rise to several belts. Even with two belts, these can hardly be identified on the circular diagram and if their number exceeds two, the problem becomes even more complicated. That is why fault

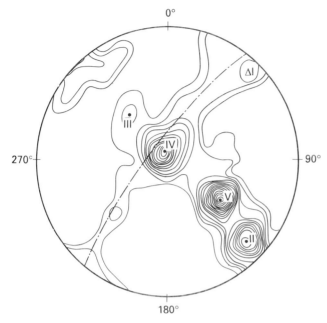

Figure 2.8 Circular diagram of fracturing with a belt of fault-line fractures shown as a dot-and-dash line. Daregor coal field (Iran), Jurassic coal-bearing formation: I – bedding joints; II–V – tectonic fold fractures. Fault-line fractures are associated with bedding-plane movement

feathering zones often feature fractures of different direction or a chaotic arrangement.

Density of fault-line fractures

Feathering joints have been studied by the authors in the Kerman coal deposit, Iran, in the Tajik depression and Mongolia with M. V. Rats and M. I. Pogrebisky, and in Dagestan limestones. Almost contemporaneously with Stoyanov (1977) and Tchalenko (1970) and separately from them, the arrangement of the five principal sets of fault-line fractures forming the feathering belt has been studied. The authors were apparently the first to study general rules governing variations in density of fault-line fractures (Pogrebisky, Rats and Chernyshev, 1971).

The fracture density trend was analysed in Meso–Cenozoic sedimentary rocks in the Tajik depression and in Upper Paleozoic granitoids of the South-Gissar structural and facies zone bordering on the Tajik depression in the north. The spacing between parallel fractures (a) was measured at points equidistant from the fault. The distance x was 10–15 km for major faults. Arrangement of the points was perpendicular to the fault strike.

Measurements were taken for tectonic faults of different size ranging from the largest regional Vaksh strike slip-cum-thrust fault, at the junction of the South Gissar region and the Tajik depression, to some minor faults extending for a few hundred metres.

After analysing fracture density versus fault distance graphs, fractures become more closely spaced near the fault (Figure 2.9). However, this relationship, if studied quantitatively, displays another feature, i.e. non-linearity. The spacing between fractures as these approach the fault decreases at an accelerated pace. This relationship can be approximated by the function

$$\bar{a} = a_b - c_{exp}(-x/k)$$

where

a_b, c and k are parameters assessed by the method of least squares. The geological content of the parameters is as follows:

a_b is background spacing between fractures in the region free from the fault effect;

$(a_c - c)$ is the limiting minimum spacing between fractures in the fault zone;

k characterizes the width of a fault-line fracturing zone. Concrete parameters of the equations for the four faults are tabulated in Table 2.6.

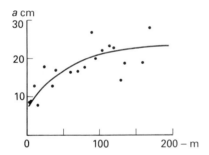

Figure 2.9 Variations in average fracture spacing as a fault is approached. Tajik depression, Vakhsh left bank, Rogunskaya Hydro; a is fracture spacing, L distance from fault

The a_b and $a_c - c$ values proved to be close to each other for different faults in the region. Background spacing, a_b, characterizes regional fractur-ing and, ideally, could be constant at least for lithologically similar rocks. The second characteristic, although related directly to the fault zone, is more associated with rock strength properties. This sheds light upon the nature of fault evolution and is in line with known hypotheses as to fracture growth (Tetyayev, 1940; Rats and Chernyshev, 1970).

Unlike the two above-mentioned parameters of the equation, the third parameter (k) characterizing width of a fault-line fracture zone is essen-tially connected with fault length. It varies in this case not two- or three-fold as a_b and $a_b - c$ but by more than 30 times. This parameter reaches a maximum at the longest faults. These observations are compa-

Table 2.6 Variations in fracture spacing at faults; regression equation parameters

No.	Tectonic fault	Rock type	Length (km)	Width (km)	Number of measurements (a)	Equation parameters					Assessment of relationship
						a_b (cm)	$a_b - c$ (cm)	$\dfrac{a_b - c}{a_b}$	K (m)	R (m)	
1.	Minor fault, Sai Gougel, Obishur basin, Vakhsh right-hand tributary	Sandstone	1.0	0.2	600	24.3	6.8	0.28	60	60–70	0.46
2.	Fault on the Vakhsh left bank, Nurek	Sandstone	0.8	0.2	700	38.0	13.0	0.34	50	60	0.56
3.	Gulizindan thrust fault, Obishur valley, near Kundyzamon village	Sandstone	65	0.8	1000	14.0	4.0	0.28	150	200	0.80
4.	Vakhsh thrust fault, near Khodzhaalisho, Tagikomar villages	Granite	100	4.1	1300	25.9	5.9	0.23	1700	1500–1700	0.35

tible with the theory of transformation of regional stress fields at tectonic
faults and more generally at any fracture (Griffith, 1921, 1924)

Of interest are the constant $(a_b - c)/a_b$ values for the Rogunskaya
Hydro area (item numbers 1, 3, 4 in Table 2.6). Individually a_b and
$(a_b - c)$ somehow characterize rock strength in the fault-line zone and the
nature of the tectonic process. In dimensionless terms, the relationship is a
constant characterizing the extent to which featuring has changed in the
fault-line zone as compared with a background.

Thus, general rules governing the structural pattern of tectonic fracture
systems have been ascertained when studying tectonic fractures in sedimentary and igneous rocks. Fold fractures form a group of fracture sets
interrelated in terms of their arrangement. The location of fold fracture
sets is determined both by that of the limb and by that of the fold axis.
Fault-line fractures give rise to feathering belts. Fault-line fracture density
patterns may be summarized as follows: as the fault is approached,
fractures become more closely spaced at an accelerated pace following a
non-linear law; major fault zones differ from their minor counterparts
mostly in width, whereas fracture density in the immediate vicinity of the
fault depends only to a small degree on fault length.

2.4   Hypergene fractures

Hypergene fractures develop at the contact of a rock mass with the
atmosphere and hydrosphere. As to their occurrence, they can be subdivided into fractures originating close to engineering objects and related
thereto and fractures brought about by natural factors.

Release fractures

Outcropping rock masses bear the traits of stress existing at depth. The
stress field in such rock masses is rearranged with relaxing stress. If the
stress relaxation rate is lower as compared with that of the mass erosional
release of a rock, rocks possessing a store of elastic energy occur close to
the surface. It is there that release fractures may develop.

Release is primarily responsible for changing fracturing parameters in a
set more or less parallel to the Earth's surface. The changes often affect
only fracture width and leave other fracturing parameters untouched. The
width of some fractures may reach 1 m (Ust-Ilimskaya and Toktogulskaya
Hydros). Fractures may not only expand, but also become denser in a set
parallel to the ground surface (Kieslinger, 1958; Bondarik, 1959; Terzaghi,
1962; Müller, 1963; Prochukhan, 1964). Variations in fracture width are
similar in nature to those in fracture spacing: as fractures approach the
surface, their expansion and density increase following the exponential law
(fracture width (Chernyshev, 1965)) and the hyperbolic law (fracture
spacing (Prochukhan, 1964)). It is not only average fracture width, but also
width dispersion that grows as the Earth's surface is approached. An
increase in fracture width dispersion is observed, in particular, in traps[*] at

---

[*] TRAP is a generic term for any dark-coloured fine grained non-granitic hypabyssal or
extrusive rock, such as basalt.

the Ust-Ilimskaya Hydro. The mutual growth of average fracture width and width dispersion shows that the release process features selectivity and inheritance, as do most geological processes. Release revives fractures inherited from earlier deformations. Fractures which were among the first to become revived expand further. Consequently, the dispersion of fracture width grows. Permeability inhomogeneities of a fractured rock mass intensify in the course of release (Sokolov, 1962).

Variations in fracture width and density in systems not parallel to the Earth's surface are negligible. Variations in width b and fracture spacing a in sets in Carboniferous limestones at Toktogulskaya Hydro (data furnished by A. V. Kolichko and M. V. Rats) are tabulated in Table 2.7. Set 1 is parallel, whereas sets 2 and 3 are not parallel to the ground surface.

Table 2.7 Depth-dependent variations in fracture spacing and width for different sets

Depth from adit mouth (m)	1		2		3	
	a (cm)	b (mm)	a (cm)	b (mm)	a (cm)	b (mm)
0.5–5.0	15	0.8	6.0	0.2	16	0.4
5.0–10.0	14	0.3	6.0	0.2	16	0.4
10.0–20.0	18	0.2	8.5	0.2	11	0.1
20.0–30.0	–	–	6.0	0.1	17	0.1

The release process generally does not give rise to a new fracture set parallel to the Earth's surface. The following conditions are presumably essential for a horizontal fracture set to originate: (i) the rock mass should have no fractures more or less parallel to the Earth's surface; (ii) the rock mass should feature fairly high horizontal compression stress making a layer unstable, which in turn may give rise to vertical tensile stress. While investigating Norwegian granites, Kieslinger (1958) described release fractures with new fractures being formed if the existing ones intersect the Earth's surface at angles exceeding 20°.

The term 'release fractures' applies not only to those fractures which have originated in the release zone, but also to earlier fractures essentially transformed in the course of time. The degree of transformation under which a tectonic or petrogenetic fracture may be termed 'release fracture' is, of course, taken conventionally. Yet, there are a few morphological features which, if taken together, enable us to attribute fractures to release. First of all these include their proximity to the Earth's surface and their width of several centimetres or decimetres, their strike more or less parallel to the Earth's surface, and gaping or infilling with eluvium or even Quaternary deposits of other origin.

Table 2.8, drawn up on the basis of percolation test data, shows the depth to which rock mass deformation extends following stress release. The thickness of a release zone on the Russian and East-Siberian platform is within a few dozen metres. It might be interesting to note that the lower boundary of a release zone is always conventional since fracture opening as variations of other parameters is depth-dependent and gradually reaches zero, asymptotically approaching a certain average typical of deep-seated portions of the rock mass.

Table 2.8 Thickness of release zone under different conditions

Area	Rock	Release zone thickness (m)
Dneprovskaya Hydro	Granite	15
Kremenchug Hydro	Granite	15
Ukraine	Granite gneiss	30
Bashkiria	Limestone	30
Bratsk Hydro	Trap	30
Ust-Ilim Hydro	Trap	30
Krasnoyarsk Hydro	Granite	25
Bratsky Hydro	Ordovician sandstones and siltstones	30
Ust-Ilim Hydro	Carboniferous sandstones and siltstones	50

The mechanism by which technogenic release and weathering fractures are formed is similar to the one for natural fractures. They, too, feature inherited evolution.

Weathering fractures

Weathering greatly affects rock fracture systems (Figure 2.10). It results in expansion of existing fractures and gives rise to new weathering fractures arranged, as a rule, in a particular way. Weathering, much like release, features selective and inherited evolution. Rock mass heterogeneities intensify in the course of weathering. On the one hand, a zone of weathering has very wide fractures (see Figure 1.9) filled with clayey weathering products, whereas on the other hand, side-by-side with those are unweathered rock blocks not affected by original and tectonic fractures and hence remaining intact in the course of release and weathering. Weathering spreads from the Earth's surface first along major fractures, then along minor fractures.

Studying rock transformation in a zone of weathering, Kolomensky (1952) and Zolotarev (1962) singled out four levels or zones, namely (i) solid; (ii) blocky; (iii) microclastic; (iv) finely crushed. This classification is fairly common in the USSR. Following Ruxton and Berry (1957) and subsequent modifications (Anon., 1981a), Western schemes for description of a typical weathering profile tend to be more detailed, with three or five divisions. In the simplest version the following zones are recognized from the surface downwards:

soil	extremely or 'completely' weathered
soil and rock	moderately to highly weathered
rock	fresh to slightly weathered

Weathering implies either chemical or mechanical weathering, or a combination of the two.

Figure 2.10 Fracturing in a zone of basalt weathering, outcrop 8 m high (Orkhon valley, Mongolia)

The principal features of weathering fractures, both newly formed and those evolving from earlier fractures, are as follows:

- proximity to the Earth's surface;
- weathering of the walls;
- walls stained, and/or the presence of a filler;
- small length;
- varied arrangement.

The last two features make weathering fractures different from release fractures.

Blast fractures

Blast fractures, as other hypergene fractures, evolve essentially by inheriting surfaces of weakness already existing in a rock mass. Following the blast, rock fractures usually expand and become more closely spaced. If rocks are strong enough, it is mostly earlier fractures that are revived (Kolichko, 1966). Rock is crushed only in a zone adjacent to the charge (Figure 2.11).

Blast fractures may be of great engineering and geological significance. According to Macovec (1962), blasting operations performed in the foundation excavation of a hydroelectric power station on the Danube resulted in a 1.8–3.5-fold increase in water permeability in a 2 m-thick zone.

Figure 2.11 Blast fractures, more closely spaced as charge loading points are approached, and blast-revived tectonic shear fractures (rectilinear) in Lower Cretaceous limestones in Kara–Koisu valley; outcrop 3.5 m high

Summary

Consider once again the principal features of hypergene fractures. With a rock mass subjected to hypergenesis, fracture width is the first to be affected. Next a fracture filler is transformed and fracture wall rocks are weathered. Further changes lead to greater density of fractures in existing sets and, lastly, fractures change their arrangement to that which is most stable in hypergenesis. This indicates that in the hypergenesis zone fractures originate mostly in already existing sets.

Another principal feature of hypergene fractures is an increase in fracture width dispersion and other parameters (Chernyshev, 1965). In a zone of weathering, dispersion increases from the bottom upwards, which reflects its increase with time in the course of weathering.

2.5   Genesis-dependent fracture system symmetry

The most general rules governing the arrangement of fracture systems can be accounted for by symmetry, and four classes have been singled out (see Table 1.1). Of these, three classes were empirically singled out previously (Rats and Chernyshev, 1970). It is well known that fractures are symmetric about the axes of principal normal stresses following distintegration of rock as a continuous medium. A pair of shear fracture sets is bilaterally symmetric about the axis of minor principal normal stress. The above fracture sets intersect along the axis of intermediate principal normal stress

(δ_2). Tension fractures are perpendicular and parallel to the axis of major principal normal stress. These fracture systems are symmetrical and reflect stress tensor symmetry.

With fractures of some particular generation growing, stresses are constant in individual lithologically homogeneous and structurally isolated parts of the rock mass, in other words in each 'domain'. It is there that a fracture set is characterized by a particular group of symmetry elements. Rock mass heterogeneities lead to stress fluctuation, brittle failure and consequently to a statistical expression of symmetry.

Each generation of fractures in a rock mass reflects the stress field existing at the moment of fracturing. Stress field changes give rise to new generations of fractures. Each generation of fractures has its own group symmetry. With generations superimposed on each other, their dissymetry is combined. The number of automorphic transformations in a system decreases and it gradually becomes ever more complex and loses symmetry.

Due to stress fluctuation and combination of fracture generations, fracture system symmetry gradually becomes more and more complex.

The above-mentioned major genetic types of fractures are the relicts of particular stresses. These types of stresses and ensuing fracture systems can be analysed, assuming that before a new generation of fractures is born, all the previous ones are healed. Critical stresses arise at some material point of a continuous medium at a moment of fracturing.

At the first stage during petrogenesis and lithogenes is, tensile stresses originate, directed towards all sides of the point in question with the same intensity, which gives rise to a spherical stress tensor in isotropic rock. Contraction (rock tension) is usually combined with gravity pressure. Net contraction is actually rare in cases where weight is negligible as compared with contraction force. A spherical stress tensor has been found in shell-like jointing of spheroidal lavas and concretions (see Figure 1.4). Such spheroidal jointing results from weathering of some isotropic rocks such as siltstones, granites (Figure 2.12) and also where a rock block develops radially directed temperature compression and tensile stresses for exceeding gravity stresses.

Combination of contraction and gravity forces make principal normal stresses unequal. Vertically, tension is compensated for by compression. Stress dissymmetries are combined. A polar vertical axis of symmetry and the infinitely large number of non-polar horizontal axes of symmetry remain common to the stresses. The growth of tension fractures is parallel or perpendicular to the axes. The symmetric (statistically) polygonal joint columns define the position of the vertical stress symmetry axis. Platy jointing, also polygonal in plan(see Figure 2.1), originates in a laminated sediment with a large number of boundaries.

The second type of fracture systems (see II, Table 1.1) is more common. Polygonal fracture systems of practical importance are mostly found in basalt effusives (Figure 2.13). Polygonal jointing is formed in an isotropic rock mass and may also develop in a transversely isotropic rock mass when the axis of anisotropy coincides with the vertical axis of principal normal stresses. The term 'transversely isotropic rock mass' implies any sediment-ary unit with horizontal bedding.

Figure 2.12 Spheroidal jointing in granite boulder resulting from weathering, Kolyma

Figure 2.13 Polygonal system of fractures in basalts, Armenia (photograph by Kh. G. Khachaturian)

Tectonic stresses arise in a rock mass from directional external compression or, more rarely, tension. The direction of principal normal tectonic stresses usually does not coincide with anisotropy axes, which leads to inequality of all the three principal normal stresses. These axes become axes of symmetry in a fracture system being formed. There originate tension fractures coinciding with planes of the Cartesian coordinate system plotted on the axes of principal normal stresses. Shear fractures following the direction of major tangential stresses are also formed.

It should be noted that there is a finite number of directions along which fractures develop with the tectonic stress field being constant. The position of axes of principal normal stresses is determined as the symmetry axes in a fracture system.

The finite number of directions is realized by the appropriate number of fracture systems. Thus, the third regular type of fracture system is formed. If the axis of major principal normal stress coincides with that of anisotropy of a transversely isotropic medium under tectonic deformation, a polygonal system of tectonic fractures is formed as in the case of diapirs.

In terms of time of origin, tectonic fractures are followed by hypergene fractures. These are formed at some point of an anisotropic rock mass in the following way. Close to the Earth's surface the stress tensor rearranges under the effect of release, which generally makes principal normal stresses unequal. It may be emphasized that in terms of fracture and stress symmetry the picture is similar to that of tectonic stresses: a set of fractures develops parallel to the Earth's surface.

The discussion up till now has been focused on fracture systems, with fractures of earlier generations being either healed or numerous and small, thus enabling the rock mass to be treated as a continuous medium. Many fractures exist in a rock mass for a long period of time. Fracture generations become superimposed on each other. Fracture systems are combined, as is their dissymmetry. Gradually, a system loses its symmetrical properties and becomes asymmetrical or chaotic. It is then that the fourth type of fracture system is formed.

A group of fracture system symmetry may vary not only through disintegration of rock blocks with new generations of fractures superimposed on each other, but also under plastic deformation, say, in folds, which reduces the number of automorphic transformations, making the system almost asymmetrical.

The above types of stresses bringing about material failure make up all possible combinations of principal normal stresses σ_1, σ_2, and σ_3. Accordingly, the fracture systems these stresses are responsible for include all possible types.

The genetic aspects of formation of fracture systems with different extent of intermittence are less clear. Original fracture sets generally develop until fractures fully merge. However, at the stage of katagenesis in a sediment, when pressure–temperature conditions are much different from those of deposition, these fractures are healed partially or fully. In igneous rocks healing is related to pneumatolytic or hydrothermal processes. Tectonic fracture systems do not lead, as a rule, to the full disintegration of a rock mass into blocks. They are non-persistent and often sub-persistent. It is only a system of fault-line tectonic fractures that is often persistent. Fracture arrangement and symmetry rarely change, as noted above, during hypergenesis. By contrast, intermittence of fracture systems radically changes and all of them become persistent.

For practical purposes classification of fracture systems has been combined on the basis of symmetry and intermittence (Chernyshev, 1979).

Spheroidal fracture systems have not been taken into account since these are rare and of little practical importance. The classification (Figure 2.14) accounts for the following fracture systems:

(a) regular, polygonal and chaotic persistent;
(b) regular, polygonal and chaotic sub-persistent;
(c) regular, polygonal and chaotic non-persistent;

Regular fracture systems are most common in rock masses. Localized fracture systems occur in some engineering problems; for instance in surface civil engineering construction, regular and chaotic persistent fracture systems, typical of the hypergenesis (weathering) zone, are usually present. Underground construction involves deeper levels of the Earth's crust where there are sub-persistent, mostly regular and chaotic fracture systems typical of unweathered rocks affected by tectonic and original fracturing. Non-persistent fracture systems are the most rare. These may be present in young rocks not subject to considerable tectonic stress, and in the inner portions of batholiths.

The above-mentioned geometrical classification of fracture systems summarizes the geological and genetic evidence of fracturing. It is there that geologico-genetic and mechanico-mathematical approaches to the study of fracturing are closely interrelated. Reflecting natural features of fracture system, it provides the basis for solving engineering problems through mechanics.

Each class of fractures calls for a particular approach to be made both in field investigations and engineering calculations. For example, when

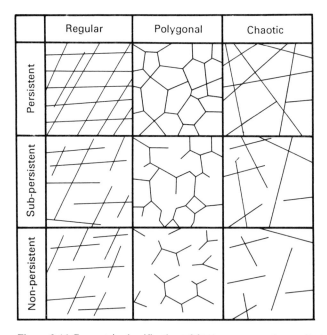

Figure 2.14 Geometric classification of fracture systems for applied purposes

assessing strength of a rock mass with a regular, persistent fracture system rock strength in a block may be disregarded, which greatly facilitates calculations. A tensor permeability theory can also be applied to this fracture system and to the regular, sub-persistent system, although in the latter case it must be corrected accordingly.

2.6 Cleavage

Cleavage is a combination of closely spaced approximately parallel surfaces along which rock splits into thin lamellae. Cleavage surfaces appear, as a rule, upon rock weathering or splitting; therefore cleavage is often referred to as the capacity of rock to split in a particular direction. As regards formal morphological features, cleavage does not differ from fractures or, more specifically, from microfissures (see Section 1.1). In terms of a rock mechanics treatment of rocks as a material, cleavage represents a surface of weakness in a rock mass, similar to fractures or layer boundaries, featuring minor dimensions of individual defects. However, the habit of this phenomenon, its relationship with mineral composition, texture and structure of rocks enable geologists to treat cleavage as a special structural form.

It is customary to differentiate between flow and fracture cleavages (Belousov, 1962). Flow cleavage is the ability of rock to split along parallel surfaces determined by grain contacts and cleavage planes of elongate mineral grains. Flow cleavage is typical of rocks containing clay minerals, micas, chlorite, hornblende, i.e. minerals with perfect and fairly perfect cleavage in a certain direction. This type of cleavage is associated with the flow of crystalline rock material under high pressures. In this case crystals are arranged with their long axes in the flow direction. The displacement of material is evidenced not only by rearrangement of lamellar minerals, but also by deformation of oolites and faunal remains.

Flow cleavage originates from compression, following the direction perpendicular to the maximum compressive force. This is confirmed by laboratory experiments performed by Daubree on clay and metal samples. Experiments by Kirillova and Chertkova (Belousov, 1962) on wax, paraffin, stearine, and ozokerite supported the hypothesis on rearrangement of crystals in the event of crushing, and coincidence of cleavage surfaces with the position of long crystal axes. Maxwell and Rozanova (Belousov, 1962), while observing cleavage in intercalated sandstones and clay shales, have shown that during flowage sand material was drawn into clay material along the cleavage surfaces. Thus, there is abundant evidence to the effect that cleavage is related to material flow. By origin, flow cleavage cannot presumably be identified as fractures since the latter result from brittle failure of rocks.

Fracture cleavage differs from flow cleavage in that (i) the direction of surfaces of weakness in a rock mass does not coincide with the arrangement of mineral grains and does not depend upon rock texture or structure; and (ii) surfaces of weakness are more clearly defined, whereas rocks open almost at any point along flow cleavage. In terms of form and arrangement of surfaces of weakness and mode of intersection of mineral grains, fracture cleavage (as the name implies) is closer to fractures than flow cleavage.

However, fracture cleavage may be so closely spaced that it can hardly be differentiated from flow cleavage without microscopic analysis. Cleavage surfaces are parallel to each other and spaced 0.1–0.2 cm apart. Some mineral grains on the surface of weakness are arranged so that their axes follow the cleavage plane direction. Fracture cleavage is morphologically related to flow cleavage in that the flow cleavage surfaces in plastic rocks acquire some features of fracture cleavage when passing to a neighbouring, more rigid, layer. Generally speaking, flow cleavage grades into fracture cleavage as pressure builds up and material becomes less ductile. Some differences notwithstanding, for engineering purposes all this makes the two principal genetic types of cleavage fairly similar to each other.

Cleavage does not always result from rock flowage. Flowage of rock salt, marble and even gneiss is possible with no cleavage formed. Cleavage is most common in clay rocks; it is rare in argillaceous limestones and marls and even less common in argillaceous and calcareous sandstones. It is interesting to note that layer thickness decreases as material is squeezed from the fold limbs into the fold hinge which points to a relationship between cleavage and plastic rock flow.

Cleavage occurs essentially in folded strata, with several geometrical types with respect to bedding and the position of the fold axial surface:

1. Basal flow cleavage parallel to the fold axial surface: it may pass from one fold to another, preserving its arrangement, and crosses layers not along the strike at periclinal fold closures.
2. Transverse cleavage develops in the plane perpendicular to the fold axis.
3. Direct fanning cleavage represents numerous surfaces which coalesce below a syncline or above an anticline and vice versa in reverse fanning cleavage.
4. Cleavage with the bedding is parallel to the layer surface and associated with a slip of layers along the bedding surfaces in the process of fold-bending.
5. S-surfaces (see Skempton, 1966, Figure 7) generally have the shape of an integral sign in cross-section. Such a form of cleavage occurs in layers with a composition gradually varying from the top to bottom or under deformation of a layer once cleavage surfaces (1–3) are formed. Deformation is often associated with the difference in rates of slip between the top and bottom of a plastic layer interposed between rigid layers. There are also some other forms of cleavage found more rarely, say fault-line cleavage, identified by Mikhailov (1956), occurring locally along the fault plane parallel to the fault surface.

Cleavage mostly intersects bedding surfaces. Goncharov (1963), studying cleavage in terrigeneous rocks of the Upper Devonian in the Urals, has shown that the angle between cleavage and bedding and cleavage surface density depend upon granulometric composition of rocks. As the average grain size decreases, the angle between cleavage and bedding also decreases, whereas cleavage density increases.

It should also be noted that, genetically, cleavage is related both to brittle failure (fracture cleavage) and to the plastic flow of mineral grains (flow or slaty cleavage) and hence cannot be entirely associated with fractures.

According to the physical theory of fractures, the latter represent cavities formed as a result of brittle failure. Cleavage, much like surfaces of weakness evolving along the surfaces of oblong grains and cleavage planes, is genetically close to schistosity. Taking into account the double nature of cleavage and the fact that this term can be fully replaced by two other terms (schistosity and fracturing), some authors consider that the term 'cleavage' should be discarded from geological use. Such a practice already exists in Soviet engineering geology where cleavage surfaces cannot actually be differentiated from microfissures. They generate anisotropy of strength, deformation and permeability properties of rocks and are studied as schistosity or fracturing. Therefore, the term 'cleavage' will not be used hereinafter.

Chapter 3
Evolution of rock fractures and fracture systems

3.1 The concept of inherited evolution of tectonic structures

There are a number of hypotheses of the evolution of certain genetic types and systems of fractures, say the formation of contraction fractures in batholiths (Cloos, 1923), frost clefts (Dostovalov, 1959; Grigorian, 1987), release fractures (Lykoshin, 1953; Kieslinger, 1958; Bondarik, 1959; Savage and Varnes, 1987), weathering fractures, and so on. All of them cover some specific environment of fracture formation and specify features of formation of individual genetic types of fractures. A fracture system represents, among other things, a combination of fractures of different age and origin. Many fractures are revived at different stages of the geological history of a particular rock mass. They can hardly be attributed to any particular genetic type. That is why it might be interesting to develop a general theory of rock fracture evolution. Such a theory will be essential for engineering geology concerned with fracture investigations. This chapter is aimed at revealing the most general rules governing the evolution of rock fractures and their patterns which can be explained by the principle of inherited evolution of geological structures.

The principle of inheritance worked out by the Moscow school of geologists is essentially a principal law of evolution of geological structures. According to Peive, 'structural formation is a progressive process featuring two opposing principles, namely inheritance and neogenesis of structures. This is the essence of the principle of inheritance' (1956, p.30).

Inheritance is traced in the evolution of major crustal elements throughout the geological history of our planet.

When studying deep faults Peive established that 'linearity and in many cases rectilinearity of geomorphological and structural features brought about by crustal faults was not as typical of the Archaean and Early Proterozoic periods as it was of all the subsequent epochs' (1956, p.12). Linearity originated from stable structural patterns of principal deformations throughout the Phanerozoic. Numerous ancient faults in the Earth's crust are known to occur in different parts of the world. Yanshin (1951) established the inherited evolution of folds in the cover-rocks of young platforms. Inherited folds keep evolving after consolidation of the basement of the Epi-Caledonian and Epi-Hercynian platforms during 120–200 Ma. Inherited evolution involves the same deformation pattern and structural arrangement.

Speaking in more general terms 'inheritance is evolution of major structures at the expense of minor structures that are not inherited' (Yanshin, 1951, p.276). Small faults, unlike small folds, feature in a number of cases of more continuous inherited evolution (Peive, 1956, p.14).

Along with inherited features of geological structures, there are, 'superimposed structures' which do not agree with the previous tectonic pattern. Superimposed structures are a neogenesis, possibly possessing some of the principal features of old structures emphasizing the unity between inheritance and neogenesis.

Inherited evolution of fractures has been repeatedly dealt with in mining geology. Pospelov noted that 'fracture systems containing dykes and veins are not newly formed, but revived, the fractures growing in length and opening' (Lukin, Kushnarev and Chernyshev, 1955, p.25). The latter authors emphasize repetition of the directions of heterochronous fractures. Based on a study of mineralization of fractures in ore bodies, the authors single out several stages in the evolution of one and the same set. It might be interesting to note that both shear and tension fractures existed at different stages. Hence, a fracture is a complex geological formation which can be identified either with a model 'shear fracture' or with a model 'tension fracture'.

3.2 Fracture form and its transformation in tectonic movements

The form of a fracture wall is fairly complex, but that of the inner space located between the walls is much more complex, reflecting the form of the two walls at one time. Fracture wall roughness is characterized by several parameters measured with respect to the general direction of a particular fracture (Anon., 1980; Patton, 1966; Barton, 1976). These parameters are as follows: angle of inclination of the elevated portion, height of the elevated portion and wave length. The form of the elevated portions varies and depends, as noted above, on surface genesis. In practical problems elevated portions are treated as triangles or trapezoids. There are mega-irregularities with a wave length of about 10 m, mesoirregularities with a wave length of about 1 m, macroirregularities with a wave length of about 1 dm and microirregularities with a wave length of several millimetres to a few centimetres. The walls of long fractures generally feature irregularities of all levels. Angles of inclination of elevated portions reach a maximum of 40–60°, averaging from 5 to 30°.

Practically, it is more reasonable to discuss the form of fracture inner space. The distance between the fracture walls is actually ever changing. Fracture wall deviations may be considered as a random function of fracture length. Choosing a reference point and drawing the abscissa axis parallel to the general direction of a fracture, we may treat the fracture wall section as a plot of a stationary random function featuring ergodicity. The second wall provides a different realization of the same random function, the difference corresponding to the distance between the fracture walls or fracture opening.

Figure 3.1 Open tension fracture in granite, Kolyma Hydro foundation, one scale division 1 mm

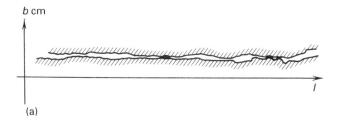

(a)

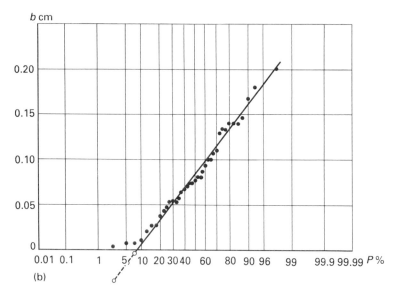

(b)

Figure 3.2 Statistical distribution of fracture width (a) graphic representation of fracture based on photograph; (b) width distribution curve

Example. Figure 3.1 illustrates a fracture section. It is not curved and its opening varies from zero to 2 mm. After measuring fracture opening in 38 sections, it may be concluded (Figure 3.2) that the distribution of opening along the length of a single fracture is almost normal (Gaussian distribution). The differences, observed only in the range of minimum values, are not great and may be attributed to measurement errors. However, there is also a non-contingent discrepancy between actual and normal distribution. Discussion of its cause helps to elucidate the actual form of a fracture and its origin. The distance between fracture walls cannot be a negative value, whereas the difference between the values of two realizations of the random function in the event of insignificant fracture opening can sometimes
be negative. A normal distribution model taken for this difference is intended, strictly speaking, for random values with no upper and lower limits and hence admits negative random values.

The normal distribution parameters in the example are such that the probability of negative fracture opening values is 5–8%. The portions where the difference of random functions describing fracture walls takes on negative values corresponds to those portions where the plots overlap (shown as drawing ink blots in Figure 3.2(a)). Actually, these portions include cut-off or squeezed fracture wall elevations. In this case the fracture wall profile reflects a process different from the one studied along 90+% of the fracture length. In particular, as is evident from the distribution curve, dispersion decreases as a result of mechanical smoothing, and opening in smoothed portions is above zero because even ground planes contact each other only on a minor portion of their surface.

Smoothed portions constitute less than 10% in the linear section of a fracture, and hence occupy less than 1% of the fracture wall area. Actually, the fracture wall contact required to transmit stresses from one block to another in a rock mass takes place over the area of some tenths and hundredths of a percent of the total fracture area. Kayakin and Ruppeneit (Ruppeneit, 1975) consider that this value is fairly constant for different rocks and amounts to 0.03%. Such a small area of fracture wall contact in a rock mass at a depth of a few tens or hundreds of metres is determined by the high hardness of minerals that compose the rocks. For example, quartz hardness is about 100 kg/mm^2, an increase of two orders as compared with the compression strength of a quartz sandstone. Calculations show that a horizontal fracture with quartz walls closes over an area of some 0.25% of the total fracture area under the pressure exerted by a rock column about 100 m high. Determination of the contact area through mineral hardness makes it possible to take account of rock mass stresses and calls for no labour-consuming measurements to be taken inasmuch as it is based on visual diagnosis of mineral in fractures and their tabulated known hardness. This problem is treated in detail in the last chapter of the book.

3.3 Statistical description of fracture sets

Actual engineering problems can be solved through mechanics on the basis of a geometrical fracture-set model only in the event of quantitative digital description of its structural pattern. There are two different methods for digital characterization, namely determinate and probabilistic–statistical. Each of them features both advantages and drawbacks. The determinate method for describing fracture sets makes it possible to characterize a local rock mass area for which fracture parameters are to be determined. However, it allows no extrapolation and interpolation due to considerable random variability of fracture systems. The probabilistic–statistical method fails to provide exact characteristics, but is capable of roughly characterizing large rock mass areas. This method enables elucidation of general rock fracture patterns so that these can be used in studying fracture origin and substantiating fracture investigation techniques.

The probabilistic–statistical approach involves the laws of distribution of fracture parameters as random values. First of all it should be noted that factual data are in agreement with the hypothesis that fracture parameters are random values. Use has been made of subseries and jump criteria (Table 3.1) to show that the above hypothesis holds true. They have also verified the hypothesis on uniform distribution of contraction and tectonic fractures in homogeneous rock masses (Rats and Chernyshev, 1970). This hypothesis supported by the results of more than 1000 measurements also holds true. Normalized autocorrelation functions have been calculated for sequences of parallel fracture spacing (Rats and Chernyshev, 1970). Each series contained 120–300 members. The normalized autocorrelation function (Figure 3.3) statistically does not differ from zero as is the case with

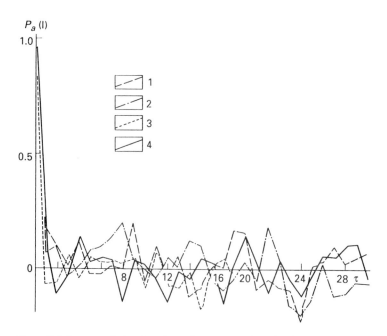

Figure 3.3 Autocorrelation functions of sequence of spacing between neighbouring fractures: (1) subhorizontal contraction fractures in traps of the Tolstomysovskaya intrusion; (2) fault-line fractures in Pateogene siltstones, Dagestan; (3) fdd-fractures in Silurian sandstones, Fergan ridge; (4) random sequence (Rats and Chernyshev, 1970)

Table 3.1 Checking random character of fracture spacing in continuous series

Outcrop location, age of rocks and genetic type of fractures	Number of measurements in series	Probability that actual data agree with hypothesis	
		criterion of the number of subseries	criterion of jump sign change
Tajik depression, Lower Cretaceous	99	0.086	0.872
sandstones and siltstones, fold	110	0.126	0.495
fractures	51	0.066	0.378
	115	0.164	0.102
	45	0.097	0.548
Tajik depression, Paleocene	24	1.000	0.402
siltstones, fold fractures	24	0.679	0.735
	17	0.181	0.682
	23	0.139	0.139
	84	0.796	0.728
	84	0.273	0.728
	154	0.705	0.897
	121	0.051	0.943
	36	1.000	0.590
	49	0.636	0.656
	35	0.387	0.682
	106	0.170	0.939
	14	0.175	1.000
South-western part of Tunguska	104	0.694	1.000
syneclize, trap sill, Triassic,	198	0.118	0.142
contraction fractures	124	0.470	0.353
	280	0.056	0.142
	94	0.097	0.162
	88	0.198	0.147
Total	2082		

random sequences. Thus, this evidence indicates convincingly that fracture spacing in a set can be treated as a random value.

The idea of uniform distribution of fractures in a rock mass and an outcrop agrees well with that of the random fracture pattern in series. This rules does not, however, apply to areas with fractures spaced ever more closely where some superimposed processes have been active.

The hypothesis on the uniform distribution of fractures was checked on the outcrop of a trap sill occurring in the foundation of the Ust-Ilimskaya Hydro. A fairly homogeneous (in terms of geology and petrography) portion of the sill was chosen with continuous mapping of fractures on a 1 : 50 scale over an area of 40 m^2. It was divided first into ten equal rectangles and then into five. Fracture trace centres were determined for one system, namely the system of horizontal fractures parallel to the sill roof. Each fracture was, therefore, represented as a point on the outcrop plane. The number of points were calculated in each square. The results of checking the hypothesis on the uniform distribution of fracture trace centres are given below.

Number of fractures

						σ	X^2	ν	P	N	$X^2_{0.05}$
4 m² site	8	10	15	6	12	10.8	22.3	9	< 0.01	108	14.68
	20	6	2	11	18						
8 m² site	28	16	17	17	30	21.6	8.5	4	0.07	108	9.49

As is evident from the table, the hypothesis of the uniform distribution of fracture traces is rejected at the 5% significance level with the area divided into ten sites. In the case of five sites, the hypothesis holds true. Upon closer consideration, the fracture distribution pattern approximates to the Poisson law.

Thus, the areal distribution of fractures should be considered to be uniform or non-uniform depending upon the area involved. This fact is of importance in engineering geology inasmuch as the foundation of an engineering structure may be considered to be homogeneous or non-homogeneous depending upon the dimensions of the structure. In the above example, trap fracturing may be considered to be homogeneous as applied to designing a dam of the Ust-Ilimskaya Hydro scheme where the foundation area is about 100 × 1000 m.

The laws of distribution of fracture parameters
The empirical solution of this problem apparently first covered the arrangement of fractures (Knoring, 1962) and the hypothesis of the log-normal distribution of fracture spacing in a system.

In rock masses subjected to no substantial deformation due to the formation of fractures, the arrangement of tectonic fractures grouped into a set may be approximated by bivariate normal distribution (Knoring, 1969). For the early fractures not grouped into sets, distribution is more complex. In situations where fractures – polygon faces – are mostly vertical, as is often the case, dip angle variations are close to normal distribution and dip azimuth variations are close to uniform distribution from 0 to 360°. Rock mass deformation results in rearrangement of fractures and the upsetting of their distribution pattern. 'Observations have shown', writes Knoring, 'that plastic deformation affects the distribution pattern of fractures in a set and of fracture sets as a whole.' (1969, p.80.) It is the symmetrical character of distribution that first becomes affected.

The Fischer circular bivariate distribution corresponds to a Gaussian distribution for unidimensional random values. Knoring has checked the actual distribution of fractures for compliance with the Fischer model distribution for a fold fracture system. The results (Tables 3.2, 3.3) show agreement with the hypothesis on the normal (Fischer) distribution of fractures within a system.

The hypothesis is not rejected at the 10% significance level and hence may be adopted. Figure 2.8 illustrates the normal character of fracture distribution within a set. The circular diagram is plotted on the basis of 200 measurements of fracture patterns in sandstones. A symmetrical fracture

Table 3.2 Azimuthal distribution of single fractures

Sector No.	Azimuthal interval (degree)	Number of fractures in interval	Average number of vectors in sector
1	0– 36	15	11.4
2	36– 72	11	11.4
3	72–108	10	11.4
4	108–144	13	11.4
5	144–180	16	11.4
6	180–216	10	11.4
7	216–252	13	11.4
8	252–288	10	11.4
9	288–324	9	11.4
10	324–360	9	11.4
	$\chi^2 = 3.72$	$\Sigma = 114$	$\Sigma = 114$

Table 3.3 Distribution of single fractures in angles of deviation from mathematical expectation

		Number of fractures in interval		
Circular area no.	Area interval (degree)	actual	theoretical	Probability
1	0– 5	19	22.58	1.00000
2	5–10	63	65.09	0.80192
3	10–15	13	10.64	0.23095
4	15–20	15	12.27	0.13761
5	> 20	4	3.42	0.02998
		$\Sigma = 114$	$\Sigma = 114$	
	$\chi^2 = 1.864$			

pattern with higher density towards the centre is observed in all the fold fracture sets from I to IV.

The linear parameters of fractures, such as length, fracture spacing in a set, amounts of relative displacement of fracture walls are interrelated. Their distribution is mostly characterized by clearly defined left asymmetry. Actual series of observations of these parameters agree well with a lognormal distribution and an exponential distribution (see Section 1.4). Comparing the two models (Figure 3.4), the distribution curves almost coincide in the right-hand portion of the graph and differ greatly from each other in the left-hand portion.

The exponential curve reflects a well-known natural characteristic of fractures related to the mechanism of their growth, namely the presence of a swarm of microfissures in rock. The smaller their length and hence the closer the spacing, the more such microfissures there are in a small rock mass. The lognormal model ignores this swarm. Series of field observations

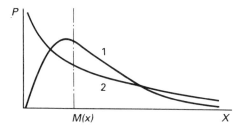

Figure 3.4 Differential curves of lognormal (1) and exponential (2) distribution of probabilities typical of fracture length, spacing and amount of relative displacement of fracture walls

agree well with the lognormal model only because microfissures are not visible to the naked eye. The drop-down distribution curve in the left-hand portion of the graph is associated with the adopted measurement system. The fact that the lognormal model selectively reflects a natural set of fractures is an advantage rather than a drawback if viewed from the angle of engineering geology. Ultraheterogeneity of a rock mass to which microfissures may be ascribed in this case is 'automatically' taken account of in laboratory studies of rock material properties. So when constructing a fracturing model of the rock mass, there is no need to transfer all fractures from the sample to the model. The lognormal model is preferable since it shows good agreement with observational data (Table 3.4).

Table 3.4 Some results of checking the hypothesis on lognormal distribution of fracture length and spacing in a set

Series no.	Rock, observation area, genetic type of fractures, measured parameter of fracturing	Number of samples	Pirson criterion χ^2	Probability P
1	Dolerites, Triassic, East-Siberian Platform, contraction fractures, fracture spacing	56	1.90	0.58
2	Same	163	5.43	0.49
3	Quartz vein, Zailiyskoye Alatau, tectonic fractures, fracture length	99	3.49	0.62
4	Limestone, Cretaceous, Tajik depression, tectonic fractures, fracture spacing	97	3.21	0.20
5	Sandstone, Cretaceous, Tajik depression, tectonic fractures, fracture spacing	80	2.56	0.47
6	Same	62	1.56	0.66
7	Same	63	2.21	0.32
8	Loam, Quaternary, Mongolia, fracture spacing	192	2.4	0.50
9	Loam, Quaternary, Mongolia, fracture length	297	7.0	0.40
Total		1009		

Note: Null hypothesis on lognormal distribution is not rejected in case P > 0.05

Disagreement between the actual distribution of linear parameters of fractures and the lognormal model is typical of lithogenetic fractures. Normal distribution has been experimentally established for polygon perimeters (Rats and Chernyshev, 1970). As is evident from the central limit theorem of the theory of chances, the sum of random values features normal distribution. One may assume that the normal distribution of parameters is due to the central limit theorem. However, the normal (Gaussian) distribution is typical of polygon sides proper, i.e. of original fracture length. Figure 3.5(a) shows two cumulative distribution curves of polygon side length. They differ slightly from straight lines, thus pointing to the Gaussian distribution of lithogenetic fracture length. The distance between the opposite sides of polygons also feature normal distribution (Figure 3.5(b)).

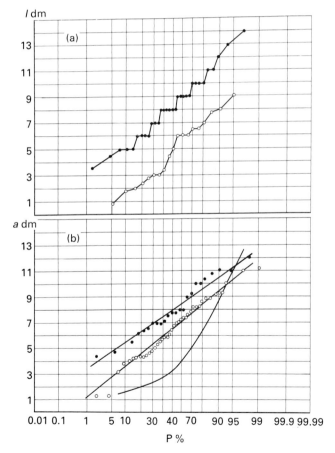

Figure 3.5 Statistical distribution of polygon side length (a) and distance between opposite sides (b) in a system of lithogenetic fractures shown in Figure 2.1 (b) The curve indicating lognormal distribution of tectonic fracture spacing is shown as a solid line for comparison.

Fracture width, one of the most important hydrogeological parameters, has long been obscure in terms of probability. Firstly, fairly exact measurements are required in this case and, secondly, fracture width is subject to change in the course of hypergenesis and engineering excavation. The character of its distribution has been ascertained only after studying the accuracy of fracture width measurements. The distribution of width of original, lithogenetic and contraction fractures does not differ statistically from normal distribution (Table 3.5).

It is likely that deep-seated tectonic fold fractures feature distribution close to normal. Hypergene and other transformations of fracture systems upset the symmetrical distribution of fracture width, making it close to lognormal. Fault-line fractures are characterized by lognormal distribution as early as in infancy. Table 3.6 illustrates the distribution of fracture width in the hypergenesis zone, where the general asymmetrical pattern is apparent.

Thus, the statistical analysis of fracture parameters has shown that these are subject to variation which is random in nature. As random values, fracture parameters feature distribution close to normal at early stages (original fracturing). In the course of evolution of fracture systems, distribution becomes asymmetrical. Fracture spacing, length and opening, as well as amount of displacement feature a left-asymmetrical distribution which may be apportioned by an exponential or lognormal curve.

3.4   Statistical description of fracture systems. The law of neogenesis of fracture sets

A fracture system in a rock mass generally comprises a number of fracture sets which originated in succession one after another. On this assumption each particular system may be treated as an individual generation. In this case a fracture system is a combination of fracture generations and ascertaining fracture system patterns is possible through study of the characteristic features of the successive fracture generations.

Studies of fracture systems as a combination of heterochronous fracture sets of different origin were undertaken long ago. They form an integral part of structural analysis as developed by Cloos. The same approach was made in tectonophysics to study fracturing of sedimentary strata (Belitsky, 1949; Kosygin, Luchitsky and Rozanov, 1949; Tchalenko, 1970). The above authors focused on general rules governing changes in the arrangement of fracture systems. Fracture length, width and spacing were considered in less detail.

Studies of fracturing have been undertaken in granites on the Kolyma, traps on the Angara, sandstones and siltstones in the Tajik depression and Iranian highland, travertines in the area of Caucasian spas and loams at the seismogenic strike slip fault in Mongolia. Factual data collected over a period of several years are summarized below. A total of about 5500 measurements of various fracture parameters have been made, from which certain conclusions may be drawn. Of special interest is the identification of successive generations of fractures in the field. Simple classical methods

Table 3.5 Checks on the hypotheses on normal and lognormal distribution of fracture width (Pogrebisky and Chernyshev, 1974)

| | | Sampling | | | | Distribution | | | | | |
| | | | | | | Normal | | | Lognormal | | |
No.	Location, rocks	Genetic type of fractures	Number n (pcs)	Average value b (cm)	Standard deviation S_b (cm)	χ^2	$\chi^2_{0.05}$	hypothesis	χ^2	$\chi^2_{0.05}$	hypothesis
1	Vakhsh River, recent sediment	Lithogenetic	154	0.75	0.28	3.95	9.48	Not rejected	212.7	7.82	Rejected
2	Kolyma Hydro, granite	Contraction	165	0.14	0.07	4.73	7.82	Same	–	–	–
3	Nurek Hydro, sandstones	Contraction Fold	108 79 54	0.14 0.06 0.05	0.07 0.03 0.03	6.11 1.47 3.40	7.82 7.82 7.82	Same Same Same	– 73.9 34.0	– 9.48 9.48	– Same Same
4	Nurek Hydro, exp. nos. 108, 109, limestones	Fold in hypergenesis zone	66 97	0.07 0.18	0.13 0.30	40.35 33.8	12.59 12.59	Rejected Rejected	7.76 10.69	12.59 12.59	Not rejected
5	Mogod–somon, seismic dislocation, loams	Fault-line	239	25.2	16.3	86.3	9.5	Same	8.4	11.1	Same

Table 3.6 Distribution of fracture width in hypergenesis zone

No.	Location, rocks	Fracture width (mm)														
		0.00 – 0.25	0.25 – 0.50	0.50 – 1.00	1.0 – 1.5	1.5 – 2.0	2.0 – 3.0	3.0 – 4.0	4.0 – 5.0	5.0 – 6.0	6.0 – 7.0	7.0 – 8.0	8.0 – 9.0	9.0 – 10.0	10 – 15	15 – 20
1	Ust-Ilim Hydro, trap intrusion, adit no.1, 0–6 m	29.5	10.5	18.9	7.4	13.7	6.3	1.1	5.3	1.0	0	2.1	0	1.1	1.0	2.2
2	Same, 6–20 m	8.5	47.5	30.5	6.5	6.5	0.5	0	0	0	0	0	0	0	0	0
3	Nurek Hydro, sandstones and siltstones, adit no.129, 0–24 m	60.0	21.8	14.1	2.3	0.9	0.3	0.3	0.3	0	0	0	0	0	0	0
4	Same 24–41 m	14.8	46.3	32.5	6.3	0	0	0	0	0	0	0	0	0	0	0
5	Same on outcrop 130 m long	48.6	15.6	21.4	6.9	4.1	2.0	0.5	0.3	0.1	0.3	0.2	0	0	0.1	0
6	All locations together	41.8	21.1	19.8	5.3	3.4	1.3	0.4	0.5	0.1	0.2	0.2	0	0.1	0.1	0.1

Note: A total of 1730 measurements were taken.

long used in structural geology may be employed for identification of successive generations of fractures.

Pollard and Aydin (1988, p.1194), in an historical review of jointing over the past century, quoted from the early literature (Woodworth, 1896; Van Hise, 1896) evidence of late shear displacements across existing fractures, presumably opened initially as joints. More recent field studies of slicken-sided earlier plumose structures on joint surfaces (Barton, 1983), and of sheared joint fillings (Nickelson and Hough, 1967; Segall and Pollard, 1983a) indicate at least two stages of deformation with markedly different strain fields. For example, development of two episodes of fracturing in granodiorites in the Sierra Nevada, USA (Segall and Pollard, 1983b), began with extensional strains accommodated by dilation of one non-persistent joint set. Later shear strain was accommodated by slip on some of the fractures of this set to produce small faults. Relic joints with undeformed filling coexist with small faults with highly deformed fillings. Splay cracks, similar to tension fractures developed in shear zones (Figure 2.7(a)), opened at the end of some of the small faults. Dyer (1893) distinguished similar sequential fractures in sandstones.

The classical geological approach does not prove to be effective on all outcrops, especially those where considerable fracture lengths are involved. Then, tectonophysical analysis has a significant role to play in two cases out of the six considered below. Changes in the stress fields and arrangement of fractures in traps (see Section 4.4) and loams in the area of seismic dislocations of the Mogod earthquake were studied (Pogrebisky and Chernyshev, 1974). Traps (Table 3.9) feature original vertical fractures; horizontal fractures comprise the second generation. Two pairs of conjugate shear fractures have been found in feather fractures along the seismogenic strike slip fault (Table 3.13), one coinciding with the fault in the first pair, and the other forming an angle of 65° with the first. Subsequent generations of fractures originate from friction on the fault plane, forming an angle of 30–35°.

Take a space vector as a fracture model (α – dip azimuth, β – dip angle, a – distance to parallel fracture, l – length, b – width). Using such an approach, an infinite fracture system may be statistically characterized by a continuous multivariate function of probability distribution. After studying the parameters of the function, it has become possible to ascertain certain rules governing changes in its form, when passing from one generation to another, which proved common to fractures of different origin.

In Table 3.8, $\bar{\beta}$, $\bar{a}$, $\bar{l}$ are arithmetic averages, s is root-mean-square deviation, v is the variation coefficient, n is number of measurements, r is assessment of correlation coefficient, and s_r is standard error of correlation coefficient.

Fractures in granites (Table 3.7) were studied in the middle part of the intrusion cropping out in the mid-course of the Kolyma river near the village of Sinegorye. Both original and rejuvenated fractures were analysed.

Original fractures in the trap sill were studied in the mid-course of the Angara near the village of Nevon (Table 3.8). To compile this table, a total of about 2000 measurements were taken. The geology of the area and fracturing are described in Section 4.4.

Evolution of rock fractures and fracture systems

Table 3.7 Parameters of contraction fractures in granite batholith, Kolyma

Location	Fracture parameters	Distribution parameters	Relative age of fracture set		
			1 old	2 middle	3 young
Outcrop	a (cm)	$\bar{a}$	73.9	107.3	195.6
		s	37.5	86.3	152.2
		v	0.5	0.8	0.8
		n	42.0	41.0	42.0
	l (m)	$\bar{l}$	8.9	8.3	3.6
		s	10.2	18.6	–
		n	23.0	34.0	10.0
		v	1.1	2.2	–
Adits nos. 100, 763, 733 and 777	$\beta°$	$\bar{\beta}$	21	80	86
		s	14	21	28
		n	185	326	23
	b (cm)	$\bar{b}$	0.20	0.20	0.16
		s	0.24	0.21	0.19
		v	1.2	1.0	1.2
		n	185.0	326.0	101.0
	$\alpha°$	$\bar{\alpha}$	–	176	87
		s	–	28	32
		n	–	326	101

Measurements listed in Table 3.9 were taken in the Vaksh river valley on the limb of a large fold several kilometres wide and a few dozen kilometres long. For Table 3.10, measurements were taken on the Pabedana Mine in the Kerman coal basin, Iran.

Fracture measurements taken in Quaternary sinter deposits on the slope of Mt Mashuk are reflected in Table 3.11, where 1, as in the two previous tables, is the set of fractures parallel to the bedding.

Table 3.12 lists parameters of feather fractures in the seismogenic strike slip fault 40 km long resulting from the Mogod earthquake on 6 January, 1977.

Observational data reflected in Tables 3.7–3.12 are summarized in Table 3.13. This table expresses the first law of fracture system evolution, namely the law of neogenesis.

Fracture sets that originate in succession in a rock mass differ from each other not only in the general direction of fractures, but also in other parameters. The differences between the first and second generations of fractures are similar to those between the second and third and so on. The main distinguishing features, when passing from older to younger sets, are as follows: (i) higher standard deviation of fracture dip and azimuth; (ii) increased fracture spacing, with higher standard deviation and variation coefficient; (iii) decreased fracture length and smaller standard deviation.

Table 3.8 Parameters of contraction fractures in trap sill, Angara

Depth from upper sill contact (m)	Fracture parameters	Distribution parameters	Relative age of fracture set	
			1 old	2 young
0–10	$\beta°$	$\bar\beta$	90	0
		s	18	21
	a (cm)	$\bar a$	27	35
10–20	$\beta°$	$\bar\beta$	90	0
		s	10	13
	a (cm)	$\bar a$	24	53
	l (m)	$\bar l$	44.2	18.0
		s	31.5	17.4
		v	0.71	0.96
		n	41.0	197.0
20–30	$\beta°$	$\bar\beta$	90	0
		s	8	13
	a (cm)	$\bar a$	18	52
30–80	$\beta°$	$\bar\beta$	90.0	0
		s	7.6	8.2
	a (cm)	$\bar a$	14	54
80–90	$\beta°$	$\bar\beta$	90	0
		s	8	13
	a (cm)	$\bar a$	12	60
90–100	$\beta°$	$\bar\beta$	90.0	0
		s	9.3	14
	a (cm)	$\bar a$	9	25

Table 3.9 Tectonic fold fracture spacing, Cretaceous siltstones, Tajik depression

Distribution parameters for a (cm)	Relative age of sets	
	1 old	2 young
$\bar a$	14.9	32.6
s	64.0	18.7
v	0.43	0.57
n	38.0	35.0

Table 3.10 Standard deviations of fracture spacing in sets, Jurassic sandstones, Iranian Highland (300 measurements)

		Relative age of set			
Station	1 old	2	3	4	5 young
95	5.8	13.6	26.4	46.4	76.4
96	–	7.7	23.6	45.2	122.5
97	–	25.3	31.6	63.1	217.0
98	–	20.6	31.8	39.1	52.2

Table 3.11 Parameters of planetary fractures in Travertines, Mt Eolovaya, Pyatigorsk

Fracture parameters	Distribution parameters	Relative age of set		
		1 old	2 middle	3 young
$\alpha°$	$\bar{\alpha}$	174.0	316.0	56.4
	s	12.8	23.7	21.4
	n	19.0	77.0	23.0
$\beta°$	$\bar{\beta}$	24.0	70.0	86.0
	s	5.7	9.2	12.8
	n	19.0	77.0	23.0
a (m)	$\bar{a}$	0.50	1.43	5.44
	s	0.23	1.28	3.24
	n	19.0	77.0	23.0
l (m)	$\bar{l}$	15.2	9.8	2.4
	s	11.7	15.3	1.8
	n	19.0	77.0	23.0

The fracture length variation coefficient shows no decrease in this case. Fracture width reflected in Tables 3.7 and 3.12 displays no general rule as regards changes in its distribution form when passing from one generation to another. This apparently stems from rapid variation of fracture width under the effect of exogenic and other factors which affect rock mass fracturing.

Considerations of the multivariate distribution function would be incomplete unless joint moments are touched upon. For this, Table 3.12 lists pair correlation coefficients which are for the most part low, not differing from zero. Of interest is the absence of relationship between fracture width and spacing on the one hand and fracture arrangement on the other hand, or between fracture width and spacing. For the youngest generation, fracture length proved to be largely determined by the arrangement of fractures within a system and by their spacing. Fracture width and length also become interrelated. This relationship is most

Table 3.12 Parameters of fault-line fractures at seismic dislocation in loams in the epicentre of the 1967 Mogod Earthquake, Mongolia

Fracture parameters	Distribution parameters	Relative age of set			
		1 old	2	3	4 young
		275.0	335.0	305.0	236.0
	s	4.4	7.9	7.8	16.5
	n	12.0	71.0	68.0	9.0
a (m)	$\bar{a}$	–	3.9	3.7	7.9
	s	–	2.6	2.9	12.1
	n	5.0	62.0	61.0	7.0
b (cm)	$\bar{b}$	19.0	19.0	23.6	12.6
	s	18.6	18.3	21.9	27.7
	n	8.0	60.0	66.0	8.0
l (m)	$\bar{l}$	55.2	18.5	22.0	7.8
	s	49.0	12.6	17.2	4.8
	n	12.0	70.0	86.0	24.0
a	r	−0.31	−0.11	−0.22	−0.57
	s_r	0.40	0.12	0.12	0.26
	n	5.00	62.0	61.0	7.00
l	r	0.08	−0.11	−0.13	−0.67
	s_r	0.33	0.12	0.12	0.17
	n	10.00	70.0	68.0	9.00
b	r	0.12	−0.10	−0.26	−0.34
	s_r	0.11	0.12	0.12	0.33
	n	8.00	69.00	66.00	8.00
al	r	−0.14	0.29	0.30	0.83
	s_r	0.43	0.12	0.12	0.12
	n	5.00	62.00	60.00	7.00
ab	r	0.26	0.00	0.17	0.24
	s_r	0.44	0.13	0.40	0.35
	n	4.00	62.00	60.00	7.00
lb	r	0.76	0.37	0.56	0.32
	s_r	0.15	0.10	0.08	0.32
	n	8.00	69.00	66.00	8.00

pronounced for fractures of the oldest generation which feature maximum average length.

The absence of interrelationship between the fracture parameters is also confirmed by the results of measurements of 120 fractures in the Lower Cretaceous limestones on the Kuli-Meyer fold in Dagestan (Caucasus). Pair correlation coefficients were calculated for such fracture parameters as dip angle, dip azimuth, length, width and spacing. Distribution of correlation coefficients was assessed and their average values determined.

Table 3.13 Changes in fracture set parameters for successive generations

Fracture parameter	Distribution parameter	Relative age of fracture set (generation)						
		1 old		2		. . .		n young
Dip azimuth $\alpha°$	$\bar{\alpha}$	$\bar{\alpha}_1$	$\neq$	$\bar{\alpha}_2$	$\neq$	. . .	$\neq$	$\bar{\alpha}_n$
	S_α	S_1	$\leqslant$	S_2	$\leqslant$	. . .	$\leqslant$	S_n
Dip angle $\beta°$	$\bar{\beta}$	$\bar{\beta}_1$	$\neq$	$\bar{\beta}_2$	$\neq$	. . .	$\neq$	$\bar{\beta}_n$
	S_β	S_1	$\leqslant$	S_2	$\leqslant$	. . .	$\leqslant$	S_n
Fracture spacing a	$\bar{a}$	a_1	$\leqslant$	$\bar{a}_2$	$\leqslant$	. . .	$\leqslant$	$\bar{a}_n$
	S_a	S_1	$\leqslant$	S_2	$\leqslant$	. . .	$\leqslant$	S_n
	V_a	V_1	$\leqslant$	V_2	$\leqslant$	. . .	$\leqslant$	V_n
Fracture length l	$\bar{l}$	l_1	$\geqslant$	l_2	$\geqslant$	. . .	$\geqslant$	l_n
	S_l	S_1	$\geqslant$	S_2	$\geqslant$	. . .	$\geqslant$	S_n
	V_l	V_1	$\leqslant$	V_2	$\leqslant$	. . .	$\leqslant$	V_n

On the whole, the study of the relationship between the parameters of tectonic fold fractures has shown that the relationship is not revealed statistically. The average values of the correlation coefficient are close to zero and the character of the distribution is close to normal, which is typical of $r = 0$. Relationship between fracture length and width, or between fracture length and spacing is an exception. Along with fault-line fractures (Table 3.12), fold fractures do not follow the rule that the longer a fracture the wider it is, or the longer a fracture the larger is the distance to a parallel fracture. Another conclusion that may be drawn stipulates that poor interrelationship between the fracture parameters allows their treatment as independent random values in a variety of engineering problems.

As is evident from Table 3.13, heterogeneities of fracture sets and systems intensify with time. This fact may be accounted for by increased heterogeneities of the stress field in a rock mass as its structural pattern becomes more complicated following the initiation of new fractures. At a certain point in the history of rock mass evolution, fracture sets are no longer formed and the regular system of fractures is replaced by a chaotic system.

There is also another trend of evolution in fracture systems. In the event of radical restructuring of a rock mass associated with plastic flow and metamorphism, the regular fracture system fades away. Structural surfaces and fractures which are formed instead have no relationship whatsoever with the earlier system of fractures. Kosygin, Luchitsky and Rozanov (1949) reproduced this phenomenon experimentally: 'Diagonal fractures are fully healed at 9140 kg/cm^2 pressure, and they cease to play a role under deformation.' (p.17.)

In conclusion, consider the engineering–geological aspect of the general rules governing changes in fracture parameters. The foregoing indicates that a system of fractures comprises several fracture sets which have different roles to play in engineering geology. Strength and permeability properties of rock masses are largely determined by the most perfect set of fractures of the first generation consisting of closely spaced long fractures following a uniform and regular pattern. Each subsequent generation of fractures makes a lesser contribution to rock mass properties. That is why dip and strike of strata are rightly regarded as a decisive engineering–geological feature of a sedimentary rock mass.

3.5   The law of inherited evolution of fractures

Consider now the evolution of an individual fracture and a combination of independent fractures in cases where a rock mass is under the influence of diverse geological factors. In the course of evolution nearly every fracture elongates, widens, coalesces or becomes healed.

Fracture growth by solution

Fracture width changes, say, at the initial stage of karstification when cavities can still be considered as fractures or slots with parallel walls.

Based on the major principles of geochemical hydrodynamics (Shestakov, 1961), and our own experiments, it can be shown that growth of a fracture is accelerated as the pressure flow of liquid dissolving its walls passes through it.

The basic equation of material transfer by seepage flow is

$$U = vC - D_{grad}C$$

The total flow rate of dissolved substance comprises convective transfer

$$U_k = vC$$

and diffusive transfer

$$U_D = -D_{grad}C$$

where

v is velocity
C is concentration
D is velocity-dependent diffusion coefficient.

Dissolution from the surface of the filter channel is:

$$\partial N/\partial t = -\gamma(C_m - C)N$$

where N is mass concentration of substance in a solid phase. The amount of substance passed into the solution is proportional to the difference between concentration of the solution (C) and saturated concentration (C_m). It is proportional to the amount of the dissolved substance in the fracture walls and the dissolution or crystallization constant (γ).

Let water flow along the short portion of a fracture with length and width denoted by l and b respectively. This flow is due to the fixed hydraulic gradient, and the flow is laminar. Water dissolves the fracture walls under conditions of convective diffusion. An experiment (Ilyin and Chernyshev, 1976) has shown that mixing of the flow in a fracture takes place within a fairly small portion of its length with $l = 10$–15 cm ($b/l = 0.001$). Assuming that the growth of concentration depends exclusively on fracture length and does not depend on its width, concentration of the solution for the fixed length l increases by p due to the dissolution of the fracture walls. The volume weight of dissolved rock is γ_o.

To determine the fracture width–time relationship: the rate of wall displacement is

$$db/dt = \Delta p q / \gamma_o S$$

where

S is the area of the two fracture walls;
q is the flow rate of water passing through the fracture;
$q = C_1 b^3$

where C_1 is constant depending upon gradient, density and viscosity of the liquid.

Under the given initial and boundary conditions, the fracture wall area is constant, as is the difference between inlet and outlet concentration (Δp) and volume weight (γ_o). Denote

$$\Delta p / \gamma_o S = C_2; \quad db/dt = C_1 C_2 b^3$$

The first derivative is known to express the rate of the process described in this case by function $b(t)$. As is evident from the last expression, the rate of fracture widening on solution increases as the cube of fracture width. Hence in cases where a large number of parallel fractures making up a set are involved, their growth is non-uniform: wide fractures develop at a much more rapid rate than narrow fractures. Integrating the expression (the rate of fracture widening), the time can be determined when the fracture width takes on some particular value:

$$b = (C_b - C_t \, t)^{-\frac{1}{2}}$$

where C_b and C_t are integration constants.

Sometimes, there is doubt that fracture widening takes place uniformly along the entire length of a fracture. The concentration of hostile carbonic acid and calcium ions in rock masses under karstification is known to be constant over a length of several kilometres. The fractures under consideration vary in length from several metres to a few tens of metres. The composition of groundwater in such fractures is almost constant, and so the rate of fracture widening is also constant. Some authors claim that mass transfer and the rate of fracture widening decrease as a fracture grows. However, these variations are minor under kinetic conditions, while under diffusive and diffusive–kinetic conditions the mass transfer coefficient varies inversely with fracture opening. In this case a variable factor $1/b$ is to be introduced into the C_1 and C_2 constants. The calculated rate of fracture

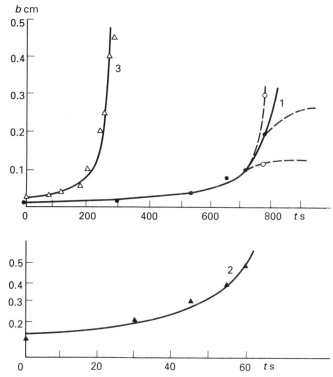

Figure 3.6 Fracture widening with time under the effect of water flow. Variations in average fracture width are shown as a solid line and those in individual fracture portions, as a dashed line

growth will decrease. Yet, the rate of widening of fractures will still exceed that of narrow fractures. This fact is important for confirming the conclusion on inherited evolution of most wide fractures.

The diffusive mechanism of dissolution should not be considered with respect to fractures with turbulent flow. Thus it may be inferred that most wide fractures hold the lead in widening under different flow conditions and with different mechanisms of dissolution. This holds true under the above boundary conditions which exist, as a rule, in natural situations where fractures are not wide. Changes in the boundary conditions decrease the flow rate of water running through a fracture with the result that its growth may be slowed down or stopped. Thus the model of fracture growth under discussion cannot be true if time tends to infinity. Fractures of infinite width are not formed due to the limited flow rate of groundwater.

To check on what has been said, $b(t)$ relations were obtained experimentally (Figure 3.6). Ice with a temperature of -0.5 °C was taken as a fractured medium and water with a temperature of $+0.3$ °C as a hostile liquid. The flow gradient remained constant ($I = 1.0$) during the experiment and the flow rate was increased to maintain the gradient. Fracture length and area were 40 cm and 552.6 cm^2 respectively. The flow was mostly turbulent. The ambient temperature was $+1\,°C$. The curves, hyperbolic in form, are approximated by the following three equations:

(a) $t = 840 - 7.6/b^{1.2}$
(b) $t = 72.5 - 5.0/b^{1.3}$
(c) $t = 280 - 3.2/b^{1.2}$

In the equations and diagram (Figure 3.6) b is fracture width in cm; t is time in seconds. With the flow splitting into jets, the observations were discontinued. Each jet was localized in some particular portion of a fracture where a tube 5–8 mm in diameter was formed. The initial stage in tube evolution is shown by curve 1. At this point the process runs under boundary conditions different from those initially specified.

An experiment with gypsum staged by James and Luptok (1978) also supported the conclusion on the priority evolution of wide fractures with their rates of widening differing greatly lengthwise.

The priority growth of wide fractures also takes place with a constant flow rate.

Fracture growth by suffosion

The groundwater flow may produce not only chemical, but also mechanical effects on fracture walls. With flow velocity v exceeding its critical value v_k, a loose fracture filler or rock material in the fracture walls is set in motion. Wherever rock is involved, washout of fracture walls, if any, is minor. However, slightly lithified, fractured rocks washout gives rise to cavities, and this process is known as suffosion.

Consider a portion of a single fracture with fixed length in the pressure flow producing a constant gradient at its ends, and assess the rate of fracture widening due to washout.

The flow velocity in a fracture is:

$$v = C_1 b^2$$

The mechanical transport of earth with the flow starts at a critical flow velocity v_k. The criterion of absence of fracture suffosion is:

$$v_k > C_1 b^2$$

Consequently suffosion cannot develop in a fracture with width

$$b < (v_k/C_1)^{\frac{1}{2}}$$

The flow rate of water passed through the fracture and hence evacuation of rock particles will vary, as in the case of dissolution, as the cube of fracture width.

As regards a combination of independent fractures with statistical width distribution, it should be noted that these fractures evolve in the same manner as under karstification, with the only difference that thin fractures with width

$$b < (v_k/C_1)^{\frac{1}{2}}$$

do not widen, whereas they grow, although at a very slow rate, on solution. The critical width of a fracture under actual conditions varies somewhat according to the gradient at its ends and some other factors. This makes karstification and suffosion pictures fairly close to one another. Prior to suffosion a rock mass featured normal distribution of fracture width, whereas in the process of suffosion the distribution pattern changes. The right-hand portion of the distribution curve straightens out and the distribution becomes left asymmetric. At the same time mathematical expectation and fracture width dispersion tend to increase.

In the reverse process of fracture healing, time is proportional to original fracture opening (Gershling, Litov and Lopushnyak, 1976). The proportionality factor is determined by a number of rock parameters. From these facts it transpires that thin fractures are the first to close, which intensifies left asymmetry of the distribution pattern with partial healing of fractures as is generally the case.

Fracture growth under load

For this, one may use Griffith's (1921, 1924) classical theory providing the basis for the modern concept of brittle failure. Force P is applied to a brittle plate whose length is b and width a (Figure 3.7). The plate has a throughgoing fracture whose length is $2l$, and $l << a$ and $l << b$. The following relation is ascertained between the tensile force acting on the plate and fracture length:

$$P = \lambda_1(E\gamma/l)^{\frac{1}{2}}$$

where

  E is Young's modulus;
  γ is the surface energy of a free surface unit;
  λ is a dimensionless factor determined by the Poisson coefficient.

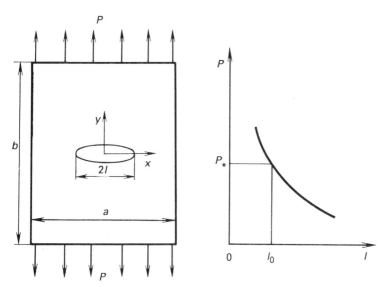

Figure 3.7 Plate with a fracture and the relation between critical tensile load and fracture length (after Griffith, 1921). P. is critical tensile load at fracture length l_0

According to the equation, a fracture does not grow when acted upon by a tensile force until the latter reaches a critical value (P_*) (Figure 3.7). At this point the fracture starts to grow rapidly and the plate breaks.

Subsequent investigations have shown that the Griffith's theory may be extended to deal with quasi-brittle failure when material displays plastic properties.

Of interest are studies in this field made by Cherepanov (1966). He dealt with the evolution of fractures in compressed bodies and ascertained the following relationship between fracture length and breaking stress under axial compression:

$$\sigma_+ = \frac{2}{(1 + P^2)^{\frac{1}{2}} - P}\left(C_f + \frac{\sqrt{(2)}L}{\pi\sqrt{(1)}}\right)$$

where

σ_+ is the compression strength of the fractured body;

φ, C_f are the angles of fracture surface friction and cohesion respectively;

L is the shear modulus of cohesion characterizing the strength of material at the fracture ends.

The foregoing proves that the relationship between fracture length and stress is also non-linear under compression, as in the case of the Griffith's theory, i.e. body strength varies inversely with the square root of fracture length. As fracture length increases, body strength drops following the same law under compression as it does under tension. With a certain fixed fracture length (l), compression strength of a body exceeds its tensile strength by

$$\frac{2C_f}{\sqrt{(1 + P^2 + P)}}$$

In the event of an ideally smooth fracture surface

$$\sigma_+ = (2L/\pi)(2/l)^{\frac{1}{2}}$$

Because of the complexity of the actual processes occurring in rock masses, the formulae of the theory of brittle failure cannot so far be applied to accurate engineering calculations. Yet the complexity of a natural medium does not impair qualitative assessment of fracture growth patterns, say, in rock burst (Avershin, Mosinets and Cherepanov, 1972; Cherepanov, 1974) or during an earthquake (Kostrov, 1974; Myachkin, 1978).

Consider a combination of fractures of different length which evolve in a rock mass under the influence of stresses. A variety of fracture lengths may be described by probabilistic distribution. At equilibrium, long fractures grow at a more rapid pace than short ones, with stresses showing a gradual and uniform increase. Non-equilibrium fractures elongating with instantaneous growth of stresses follow the rule similar to that for equilibrium fractures, i.e. in this case long fractures are the first to elongate as even a minimum increase in stresses is sufficient for this purpose. A sharp periodical increase in rock mass stresses which occurs, say, during an earthquake may be represented as a sequence of stress impulses (Figure

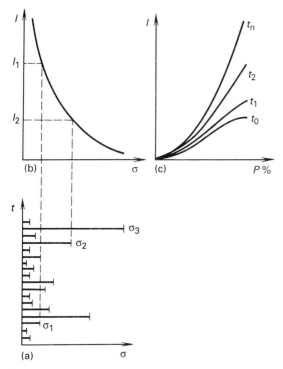

Figure 3.8 Variations in fracture length distribution with impulsive growth of non-equilibrium fractures under the conditions of regular stress increase: (a) sequence of stress impulses (σ); (b) relationship between critical length of non-equilibrium fracture and stress (after Kachanov, 1974); (c) sequence of cumulative distribution curves, from normal distribution cuve (t_0) to lognormal distribution cuve (t_n)

3.8(a)). According to the earthquake recurrence law, the greater a stress impulse, the less probable it is and the more rarely such impulses occur.

If stresses σ_1 and σ_2 are projected onto the curve $l(\sigma)$, based on the Griffith's function, values l_1 and l_2 are obtained: the length of fractures in a state of non-equilibrium increasing with impulses of stresses σ_1 and σ_2. Many fractures (see Figure 3.8(b)) whose length is less than l_1 will go in the first case into a stable state. In the second case evolution covers a vast majority of fractures. The probability of such large stresses as σ_2 is fairly small, so it is unlikely that short fractures at the level l_2 will be involved in the process of growth. The probability of elongation of shorter fractures is lower still. With a fairly long span of time, the probability of varying stresses is realized as a rate of fracture growth, which proves to be a function of the original length of fractures. In this respect transformation of the original distribution of fracture length leads to accelerated growth of long fractures whatever the mechanism of fracture growth (equilibrium or non-equilibrium). Consequently, the integral curve of fracture length distribution tends to follow the axis l passing from original normal or some other distribution to left-asymmetric lognormal distribution (Figure 3.8(c)).

The evolution of a fracture has now been considered in length and width. As for the other parameters, such as amount of displacement and spacing between parallel fractures, these are functionally related to the above parameters. Indeed, the amount of displacement is deformation accumulated on a fracture due to stepped variation of a displacement vector on the fracture surface. As is evident from Hooke's law, the accumulated deformation is proportional to a respective length. Close linear relationship between length and amount of displacement was ascertained by Cailleux (1958).

The relationship between fracture length and spacing is obvious. The growth of a fracture implies its penetration into an area between two parallel fractures, thus reducing the spacing between parallel fractures by one-half. Crushing of rock blocks is apparently consistent with Kolmogorov's hypothesis (1941), which results in lognormal distribution of fracture spacing.

Discussion

In summary the law of inherited independent evolution of rock fractures can be formulated thus: the fracture parameter x at time t_0 features normal distribution, whereas in the course of fracture evolution there arises left-asymmetric distribution at time t due to the non-linear function $x(t)$.

The priority growth of long and wide fractures, as an increase in fracture density in areas where it is high, expresses the well-known geological law of inherited evolution. The aforesaid physical mechanism accounts for left-asymmetric probabilistic distribution of fracture parameters. Yet the fracture evolution model in question has a great disadvantage since it is at variance with facts in the range of large extreme values of fracture parameters. The accelerated growth of a maximum fracture results in the fact that it approaches infinity at a rapid rate, which is obviously impossible due to the restricted geological space wherein a fracture evolves.

Consequently, there is a certain natural mechanism quite the opposite to the one of inherited evolution which makes it impossible for a fracture to evolve in full compliance with the aforesaid mechanism of geological inheritance. The tendency for inherited evolution is clearly defined by parameters of an infant microfracture, whose occurrence, trend and dimensions determine respective parameters of a fracture. Variations from the tendency for inherited evolution are caused by environmental changes, namely a geological boundary impairing fracture growth, or changes in the stress field or groundwater flow under the influence of neighbouring fractures. Many examples may be cited to show that the presence of a geological boundary adversely affects the tendency for inherited evolution.

3.6   Laws of interaction. Fracture set heterogeneity levels

Accelerated inherited evolution of fractures is observed in the case where there is no interaction between them. Interaction of fractures may yield two opposite results, namely (i) a fracture ceases to evolve under the influence of a neighbouring larger fracture or (ii) a fracture merges with a neighbouring fracture entering a new stage in its evolution; in other words, two fractures merging together give rise to a new fracture. In the first case energy required for the fracture to grow is drained by the neighbouring larger fracture. Upon the end of the fracture where stresses are concentrated reaching the other fracture, its growth discontinues. Fractures which approach one another fail to grow. This fact is exemplified by the growth of original polygonal joints. As such a joint approaches its older counterpart, it finds the shortest way to reach it giving rise to triple intersections typical of original sets (see Figure 2.1). Discontinuance of growth of original fractures, should they meet, results from localization of energy sources within polyhedrons. Energy stored by a sediment or a melt is released when tension fractures are formed. Tensile stress is removed in the vicinity of any tension fracture in the direction perpendicular to its plane. It is only in the direction along the fracture plane that tension occurs. Therefore, a second fracture while approaching a first one, changes its direction tending to intersect the first fracture at right angles. After the fractures meet, the stresses are removed along the two mutually perpendicular planes. The growth of the fractures discontinues, as is the case with approximately orthogonal fractures.

Another possibility, that of accelerated growth, is realized in the meeting of approximately parallel fractures. Sets of tectonic fractures and those transformed during hypergenesis feature varied arrangement patterns, with fractures both dying out and originating. The growth of tectonic faults due to coalescence of large fractures is a well-known fact. It should be noted that, in realizing these possibilities, all fractures fall into two groups. The larger group comprises fractures which cease to grow. The small group is represented by a new generation of large fractures which on merging become twice as long as those of the previous generation and continue to grow at a much more rapid rate inheriting accelerated growth from the merged large fractures. Thus, there arise two groups of fractures of a different order of magnitude. Statistically it is expressed by heterogenic double-peaked distribution of fracture parameters. At a certain point the growth of fractures of a new generation again leads to their coalescence and gives rise to the two groups, one characterized by accelerated growth and the other dying away. This periodically recurring process generates a number of heterogeneity levels in a fracture system, which is expressed by polymodal distribution of its parameters (Figure 3.9). According to the law of interaction, a rock mass with a clearly defined system of fractures comprises microfissures, fractures and faults. Variations in fracture length, when passing from one level to another, apparently follow the law close to that of geometric progression with base 2 or, to be more precise, a little more than 2 due to the growth of fractures at each stage before they split into the two groups. Thus, the distance between the modes in the diagram shows accelerated increase, whereas the number of members in each new leading group decreases when changing over to new levels. Data furnished by Wittke (1984) on the Barrage de Selengue (Mali) point to the presence of fractures of three levels. Fracture length is 2.5–3.0 m (level I), 9.5 m (level II) and 13.5–15.5 m (level III). Levels I, II, and III account for 95, 4 and less than 1% of fractures respectively. Accordingly, short fractures are

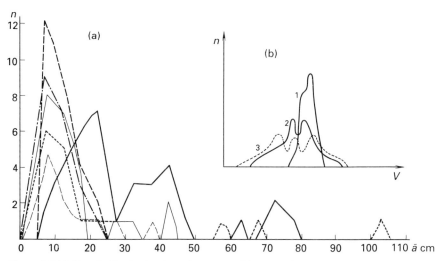

Figure 3.9 Multipeak distribution in rock masses with several levels of fracture system heterogeneity: (a) distribution of fracture spacing in limestones, Toktogul Hydro construction site (after Kolichko and Rats, 1966); (b) distribution of elastic wave velocities in traps, Talnakh deposit (after Golodkovskaya and Shaumian, 1975) for rock masses: 1 – with original fracturing; 2 – with poor tectonic fracturing; 3 – with zones of intense tectonic crush against the background of tectonic fracturing

most common, generally spaced 1 or 2 m apart. This information agrees with the hypothesis that there is geometric progression with base 2 and a little above.

Several levels of fracture set heterogeneity were clearly defined when studying the engineering geological environment of the Talnakh deposit. The diagrams (Figure 3.9(b)) show a single-peaked distribution of velocity of longitudinal waves in porphyritic basalts with original fracturing. Rock masses with poor tectonic fracturing feature double-peaked distribution in accordance with two heterogeneity levels. The evolution of tectonic fracturing results in the third peak on the distribution curve. As fracturing evolves, velocity dispersion grows. New peaks arise in the range of low velocities, i.e. they are related to areas with closely spaced fractures. The actual presence of several heterogeneity levels in a rock mass is evident from the form of curves of elastic wave velocities. In the event of a rock mass with original fracturing, the curve fades away, whereas for a rock mass with tectonic fracturing it has several bends.

The methodological conclusion drawn from the theory of inherited evolution of fractures in terms of engineering geology is as follows. The largest fractures evolve most rapidly in any process. To study fracturing, one may single out three categories, namely microfissures which are localized in rock samples and are not covered by detailed investigations; macrofractures or fractures which exist in a rock mass in large quantities and are sampled for statistical analysis; and megafractures which are not numerous but have an active role to play in different processes and hence each of them is worthy of consideration.

3.7 Salient features of structural patterns and stages in the evolution of fracture systems

There are three stages in the evolution of fracture systems. The first stage involves uniform fracturing of rocks, the fractures originating throughout a solid rock mass independently and simultaneously. Hence their distribution proves to be random following the Poisson law, whereas their arrangement is regular in compliance with that of the axes of principal normal stresses. The distribution of most fracture parameters is close to normal. Fracture spacing varies as does the length of fractures. The first stage is generally characterized by internal sources of stress fields responsible for original fracturing. However, internal energy sources may also be observed.

At the second stage fractures tend to become more closely spaced and merge. This stage features external sources of stress fields, for example tectonic. It is mostly long fractures that grow. Fractures spaced most closely show a tendency towards density, reflecting the law of inherited evolution of geological structures. Merging of closely spaced fractures is a decisive factor in the formation of tectonic faults.

At this second stage the uniform distribution of fractures becomes affected in a statistical sense because of priority growth of closely spaced fractures. The distribution of fracture parameters becomes left asymmetric through the priority growth of longer fractures. Thus, the second stage features two levels of fracture set heterogeneity. Not only do sub-parallel fractures, but also others, in particular perpendicular fractures, meet and converge. It might be interesting to note that the growth of systems of rock fractures at this stage is similar to that of metal fractures prior to failure. In both cases there is an increase in heterogeneity of the medium in a zone of anticipated fracture, localization of deformation and heterogeneous distribution of fracture parameters.

The third stage involves growth and merging of faults and transformation of macrofracture systems. Faults feature accelerated evolution. On merging, they give rise to a new level of fracture set heterogeneity, namely major faults. At the same time fault-outlined blocks appear in a rock mass. A system of faults is sub-persistent or even persistent at this stage. Small fracture systems evolve under the effect of energy released in fault zones, and also of other energy sources. Small fractures elongate and merge, thus giving rise to persistent fracture sets which dissect a rock mass into blocks.

At this third stage a rock mass is not a solid body. Dissected into blocks and having fractures infilled by rock crushing or weathering products, it may be subjected to pseudoplastic deformation due to relative creep displacement of blocks. Displaced at a slow rate, some blocks develop high stresses, giving rise to new fractures of short length and varied arrangement. These fractures obliterate regular patterns of small fractures. At this stage fracture systems become chaotic. Then comes either disintegration of the rock mass, and hence of fractures in the process of denudation, or fractures are healed in the process of metamorphism. Once fractures are healed and a rock mass regains its strength, a new system of fractures superimposed on the old system starts to evolve from the first stage.

It is large fractures that mostly grow in a particular set. Yet the tendency for inherited evolution may sometimes be adversely affected by some environmental changes. It is then that newly formed fracture sets originate. Sometimes under metamorphism, hydrothermal healing or weathering the inherited evolution discontinues entirely and a fracture system ceases to exist. Subsequently, a superimposed fracture system is formed, its parameters not being related to those of the old system of fractures. The old system can be traced as veins and dykes which are not defects in a mechanical sense and hence do not affect the course of evolution of the superimposed system.

The tendency for inherited evolution of an individual large fracture usually becomes affected in cases where it meets with another fracture. If the arrangement of these fractures is entirely different and they meet almost at right-angles to each other, one of them usually ceases to grow. Should their arrangement be similar, they merge to give rise to a new larger fracture.

Statistically, inherited evolution of fractures is expressed by different forms of distribution of most of their parameters. Because of the priority growth of large fractures, distribution of such fracture parameters as width, length and spacing through geological time becomes asymmetric with a changeover from normal to lognormal. Newly formed fracture sets differ from their older counterparts in shorter length and wider spacing, and also in greater variation of all parameters, including arrangement.

The mechanism of inherited evolution of fractures is revealed in the physical analysis of such processes as brittle failure, solution, grouting or sealing which lead to restructuring of fracture systems. All these processes have one thing in common, namely non-linearity of respective mathematical functions. Any fracture, irrespective of the process of growth, evolves under stable geological conditions in an accelerated manner. Large fractures in a group of simultaneously growing fractures evolve at a more rapid rate. Thus, the accelerated growth of fractures under stable geological conditions, which is expressed in non-linear equations describing the processes of growth, is the chief cause of inherited evolution of fracture systems.

Two important practical conclusions may be drawn in this respect. Engineering structures can be treated as neo-genesis in the Earth's crust. If this is the case, there should be deformation inherited from underlying rock in accordance with the inheritance principle. Hence it is necessary: (i) to allow for some deformation in a structure inherited from the underlying rock; (ii) to make the best use of natural structures and inherited processes in construction.

The trend of engineering and geological rock properties arises from localization of material and energy sources in a rock mass. Fracture parameters are no exception to this rule. Generally, if the process of fracturing starts from a surface which is acted upon by some forces, dispersion of fracture parameters usually decreases as we move away from this surface. The trend of fracture parameters involves a similar trend of engineering and geological rock properties. For instance, if the processes governing permeability of a rock mass start from some surface, dispersion of permeability parameters usually decreases as we move away from this surface. Variations in dispersion result from the selectivity of geological processes and inheritance in the evolution of rock mass heterogeneities.

Chapter 4
Fractures in basic igneous rocks

4.1 The formational approach to fracture study

A number of main rock parameters may be specified which determine the pattern of fractures. These include rock strength during fracturing, original mode of occurrence, type of contact with surrounding rocks (sharp or gradational), nature of the geological environment (anisotropic or isotropic) affected by fractures, and rock texture and structure. All these rock characteristics are most fully incorporated in the notion of a 'formation'.[*] It is, therefore, advisable to describe fractures in petrographic and lithological types of rocks formationwise. Fractures in some common rock formations were first dealt with even before the notion of an engineering–geological formation came into being, for instance by Cloos (1923), who described fractures in granite batholiths.

Original fractures in igneous and sedimentary rocks develop on account of internal energy stored by a melt or a sediment. The conditions of fracture growth, just as those of petro- and lithogenesis, are determined by heat and mass transfer between a particular rock being formed in the Earth's crust and its environment. That is why structural patterns of original fractures are closely related to a particular facies type. Contraction fracture patterns are governed by a respective tectonic regime, as well as by whether solidification of the melt took place at depth or on the surface. As for lithogenetic fractures, these are determined by such factors as rock composition, thickness, type of contact, and mode of intercalation. These features are characteristic not only of rocks, but of geological formations as well. By discussing rocks without reference to formations it is not possible to describe original fractures in great detail.

Tectonic fractures are generally superimposed on original fractures, their arrangement and size being inherited from the latter. It might be

---

[*] The term 'formation' as used in this context differs from conventional stratigraphical usage. The formation is the fundamental unit of lithostratigraphical classification; it is the only formal unit for subdividing the whole stratigraphical column all over the world into named units on the basis of lithology (Bates and Jackson, 1980). The term is also used for a lithologically distinctive mappable body of igneous or metamorphic rock.

It appears evident that, as an engineering geological term, formation may be extended to include sedimentary rocks. A formation may then be defined as a lithologically distinctive mappable body of rock or soil (in the engineering sense). The term is used without any stratigraphical implications.

The term may be compared to the term 'lithological complex' proposed for use in engineering geological mapping (Anon., 1976, p.12).

interesting to note that each formation corresponds to a particular stage of the geotectonic cycle and to a particular geotectonic zone (Belousov, 1962). The character of subsequent rock deformations is, therefore, somewhat predetermined by the fact that these rocks can be attributed to a particular formation. For example, the flysch formation is involved in orogenesis resulting in multiple irregular folding. The evolution of a platform carbonate formation is quite different: generally, it experiences neither downwarping nor lateral compression, but is slightly deformed as a result of vertical movements of the continental platform giving rise to placanticlines. This is also true of the porphyry (after Peive) or basalt–andesite–liparite (after Kuznetsov) formation characterized by the orogenic stage of geosynclinal development and postvolcanic rock alteration. Quite different is the history of a basalt formation evolving on cessation of volcanism at the end of an orogenic cycle. Thus the fact that we can attribute rocks to a particular formation has a strong impact on the development of tectonic fractures both mechanically through the compositional and structural features of the formation, and historically through the specific conditions of tectonic loads and deformation.

Hypergene fractures develop in rocks under the action of solar energy differently transformed on the crustal surface. Their structural pattern is determined by physico-geological conditions and landforms, as well as by the composition and texture of the rock mass. Inheriting features of older fractures, hypergene fractures also depend on a respective rock formation. It is, therefore, possible to agree with M. V. Rats who suggested that 'each sedimentary [and not only sedimentary – S. N. Chernyshev] formation is characterized by an assortment of different genetic types of fractures: original, tectonic, and hypergene, whose evolution is respectively fully, largely or partially determined by a particular formation'.

4.2 Basalt fractures

Basalt is one of the most widespread rocks. There are two basalt formations: oceanic and continental. Being much alike in petrography, these, however, differ greatly as far as their tectonic and physico-geographical environments are concerned, manifested, in particular, as entirely different types of jointing.

Continental basalts are characterized by columnar, blocky, platy, breccia and other types of jointing (Figures 4.1 and 4.2). Blocky jointing is typical of thin flows as well as of the upper and frontal portions of thick lava flows. The basaltic rock mass is dissected by numerous fractures arranged chaotically, which cross each other giving rise to triple intersections. The joint blocks are of complex oxygonal shape and measure 0.3–1.5 m across. Their surfaces are rough and curved (Figure 4.3), which is typical of tension fractures. The joint blocks are generally not interconnected by rock bridges. Tectonic deformations and weathering revive the existing fractures, new ones being formed in extremely small quantities, which can be accounted for, apart from continuous fracturing, by great rock strength and its resistance to weathering.

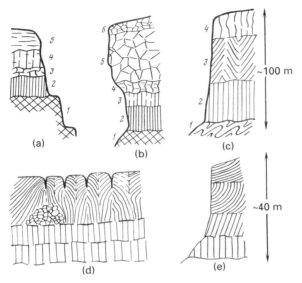

Figure 4.1 Combination of different types of jointing in lava flows (Koronovsky, 1968):
(a) Kazbek, Chekher flows: 1 – substratum, 2 – thin hexahedral prisms, 3 – blocks,
4 – thick-columnar jointing, 5 – platy jointing; (b) Elbrus, Azau flow: 1 – substratum,
2 – clearly defined columnar jointing, 3 – thick-columnar jointing, 4 – obscure columnar
jointing, 5 – blocky jointing, 6 – slag crust; (c) lavas in the Columbia valley, USA:
1 – substratum, 2 – vertical columnar jointing, 3 – inclined columnar jointing; (d),
(e) – lava flows in Elbrus, Terskol

Figure 4.2 Columnar and blocky jointing in basalts, Armenia (photograph by I. S.
Tolokonnikov)

Figure 4.3 Tension fractures in basalts, Orkhon valley, Mongolia

Columnar jointing is observed in the lower and middle portions of basaltic flows with fractures developed in a quiet geodynamic environment once the flow has stopped. The internal portions of flows 30–50 m thick display penta- and hexahedral columns, whereas thin lava flows feature columns less distinct in shape. In plateau lava sheets, the columns are rectilinear and arranged vertically. Under conditions of a dissected buried relief, the joint columns are often curved, being almost normal to the lava-bed in the lower parts of the flow and bending upwards to assume a vertical position. It is likely that in the lower part of the flow, the fractures are normal to the cooling front under the conditions of hydrostatic pressure in the internal portions of the flow. Subsequently, once these solidify, the fractures tend to become vertical, i.e. their direction is determined by gravity and by the position of the top of the flow.

In thick basaltic flows with columns several dozens of metres high, they bend along vertical fractures, which indicates that there is no difference in tensile stress in the plane normal to the column axis. Also described are fracture patterns caused by superposition of tensile stresses on contraction along the flow axis. Transverse horizontal fissures in joint columns follow no regular pattern and are spaced 0.1–5.0 m apart. The fracture wall surface is generally rough and step-like, scars and steps on the walls of vertical fractures running mostly horizontally (Pollard and Aydin, 1988,

Fig. 11). All these irregularities of the joint column result in its being pinched or thickened. Occasionally, the columns show step-like thickening downwards or upwards that fits a certain pattern.

Fracture sets in basalts of the continental formation, even outside the hypergene alteration zone, have a surprisingly high porosity ranging from 0.5 to 5% depending on rock petrography. The maximum porosity can be observed in dense doleritic basalts with columnar and blocky jointing and a minimum in vesicular tuff-lavas.

In addition to columnar and blocky jointing which are most common, basalts of the continental formation occasionally display prismatic, platy, breccia, and spheroidal jointing. Small prisms measuring 10–20 cm across are separated by thin fissures running in different directions, even within a portion a few metres long. A rock mass with prismatic jointing is generally characterized by sub-persistent fracture systems which mostly follow a fan-like pattern. Its strength is higher than that of the rock mass featuring normal columnar jointing. Basalts with breccia jointing also have sub-persistent fracture sets, fractures of irregular shape being open by about one-tenth of a millimetre. Basalts with fine breccia jointing feature maximum cohesiveness.

The exogenic alterations of fractures in basalts generally result in original fractures being either expanded or narrowed in local zones where exogenic stresses concentrate. Such zones may be related to active faults in the lava-bed, rockslides in sequences that underlie basalts, and growing salt domes. For example, basalts and underlying clays of the Eocene are cut by the Azat River in Armenia (Ter-Stepanyan and Arakelyan, 1975), the clay beds being affected by creep. A large basaltic rockslide occurred on the surface at the canyon slope. In its upper portion the slip surface intersects basalts with adjacent blocks affected by an intricate set of newly formed fractures. The original fractures in the movement zone are opened.

In the canyon of the Razdan River in Yerevan, lavas rest on a clay–gypsum unit overlying thick beds of rock salt. In this area the lava flow covered salt domes and landslides on their slopes. With the salt domes on the uprise and slow differentiated landslides occurring on their slopes, the lavas have zones of high and low permeability. Opening of fractures and higher permeability are observed in dome arches as well as over the top and marginal fractures of the buried landslides. Closing of the original fractures is visible in areas between the domes.

Basalts of the oceanic formation differ from continental basalts in that they display widespread spheroidal jointing typical of subaqueous effusives irrespective of their composition. Young basaltic lavas with spheroidal jointing were observed in situ in the Mid-Atlantic Ridge by X. Le Pichon (Riffaud and Le Pichon, 1976). Le Pichon described both spheroidal and tabular jointing of lava flows. The spheroids clearly display concentric and radial contraction fissures (Figure 4.4(a)), whereas in the case of tabular jointing, fissures run along and across the axis of the lava flow. It is interesting to note that spheroidal fragments on the ocean floor outside fracture zones are surprisingly similar in size, as is repeatedly emphasized by Le Pichon. It is likely that basalts are broken along original fractures spaced almost equally apart. In the transform fault zone, fragments of basalt in talus deposits differ in size. Along with individual large blocks

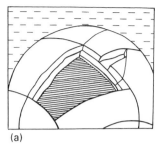

(a)

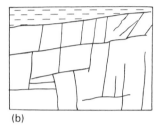

(b)

Figure 4.4 Set of fractures in basalts on the surface of a hollow spheroid (a) and on the vertical outcrop plane in the transform fault zone (b), Mid-Atlantic Ridge, depth 2.7 km (diagrams based on photographs by Le Pichon)

there occur many small fragments. The walls of the transform fault have rectilinear fissures arranged in a parallelogram pattern (Figure 4.4(b)). Thus the recent basaltic lavas on the floor of the Atlantic Ocean display characteristic features of original and tectonic fracturing.

Prismatic jointing also observed in the Atlantic basalts has presumably originated, as on the continent, in the internal portions of thick lava flows out of contact with the ocean.

4.3   Fractures in rocks of the trap formation

The trap formation is fairly common in a large number of platforms. It is regarded as one of the most widespread and well-explored igneous rock formations. Traps which are of great scientific and practical interest form sheets, sills, dykes and are represented by the tuffaceous, effusive, and intrusive facies. The dominant rock types are tuff, dolerite, gabbro, picrite, porphyrite, etc. Studies of the trap formation provide an insight into the mechanism of magmatic intrusion and extrusion under platform conditions. In this respect it is fairly important to study fracturing. Deposits of Iceland spar, graphite, iron, copper and nickel ores are generally associated with trap rocks, which makes them of great practical value. On the other hand, river valleys in areas of outcropping trap intrusions are a logical site for high dams of hydroelectric power stations. To study trap fractures is, therefore, important for engineering geology.

Fractures in rocks of the platform trap formation were described by Lebedev (1955) and studied in detail by geologists from the Gidroproyekt Research Institute in the course of designing hydroelectric power stations at Bratsk, Ust-Ilim, and Boguchany, as well as during exploitation of ore deposits for a number of projects (Mining and Concentration Complex at Norilsk, etc.). Listed below are characteristics of fractures in trap rocks as determined by these investigations.

The tuffaceous facies of Siberian traps is represented by tuff, tuff breccia, tuffite, and tuffstone alternating with typical sedimentary rocks and basalts. The rock fractures are chaotic and follow no regular pattern. The radial fractures in volcanic bombs can be found in combination with concentric fractures giving rise to shell-like jointing, which is typical of lapilli and agglomerate tuffs. The latter also include veins with longitudinal jointing. Tuffite and tuffstone are stratified, original fractures being normal to the bedding joints.

The effusive or lava facies displays columnar jointing which is typical of tholeiitic diabase and thin-columnar jointing typical of dolerite and basalt. Columnar jointing often grades into rosette jointing with fissures in long thin prismatic bodies following a radial pattern. The large-block or prismatic jointing is peculiar to coarse-grained tholeiitic and doleritic lavas, whereas small-block jointing is a feature of aphanitic amygdaloidal lavas. Spheroidal jointing is typical mostly of tuffites and tuffs. Lavas occur with spheroids 0.5–1.5 m in diameter, which apparently formed in the coastal environment. The lava-bed surface is rough, which determines the position of joint columns. The lava flows are 1.2–12.0 m thick. A relationship has been established between joint column thickness and that of lava sheets: as the latter increases, so does the spacing between joints.

The intrusive facies of traps cannot be easily distinguished from the effusive facies petrographically and hence in terms of formation conditions, the pattern of fractures in intrusive traps bears a great resemblance to that in effusives. The intrusive traps occur as stepped and multiple sills with dome-shaped swells and apophyses in the form of dykes, laccoliths, stocks.

Fractures in intrusive bodies are largely determined by their shape. Bogdanovich (1896), who was one of the first to explore traps in Siberia, wrote that intrusive traps 'display three types of jointing: columnar, which is fairly common, spheroidal or sheet. Sheet, pillow-like often of a granitic type, such as the one near the Berendinsky Bar on the Ufa River, columnar, and spheroidal types of jointing impart to rocks of one and the same geological series a peculiar look of its own.' (p.237.)

As a result of those observations, it has been possible to establish a salient feature of fractures in intrusive trap rocks, i.e. their resemblance to those in plutonic intrusives (gabbro) and effusives. Bogdanovich failed to identify blocky jointing accounted for by chaotic arrangement of fractures, which is also fairly common.

Classic columnar jointing is typical of gently dipping large bodies, such as sills and laccoliths, the columns being generally characterized by clearly defined transverse fissures. Columnar and sheet jointing is peculiar to thick lenses and laccoliths composed of a number of beds 10–15 m thick, each having vertical columnar jointing of its own. The columns are 1–40 m high and up to 8 m thick. Their thickness tends to increase as they approach the central portion of an intrusive body. Similarly, the columns are much thicker in medium- to coarse-grained dolerites than in fine-grained dolerites. Hence we can presume that column thickness is a function of rate of cooling of an igneous rock body. Thick bodies characterized by a lower rate of cooling feature larger mineral crystals and thicker joint columns. The joint columns in hypabyssal trap intrusions may be curved or inclined, as they are in effusives.

Original fractures in sills with columnar jointing differ from those in basalts at least in three aspects: (i) fractures in hypabyssal rocks are generally healed by hydrothermal minerals, such as calcite; (ii) hexagonal cross-section of columns is not typical of intrusive traps; (iii) joint blocks display fissures of the second and third generation, which is typical mostly of columns measuring 3–5 m across. Such joint columns are not found in effusives which cool at a much more rapid rate than rocks of the intrusive facies.

The columns are complex in shape, their joints, which are often curved, running for a maximum of 15 m horizontally and for dozens of metres vertically. As a rule, they are accompanied by a variety of fissures which follow no particular pattern. The length of vertical joints in a horizontal cross-section usually exceeds the cross-section of columns so that a single joint may be common to a number of neighbouring columns, which is typical of intrusive traps and found occasionally in lavas. In contrast to lavas, dolerites often show tetrahedral joint columns.

Successive generations of fractures in columns were observed by Lebedev (1955) and studied in detail by Hill (1965) in Tasmania. Joint columns are known to display radial and concentric vertical fissures which dissect the thick column with a cross-sectional area of more than 10 m^2 into smaller columns (Figure 4.5). Fissures of the second generation are thin and curved. They run along the lateral surface of columns in large quantities. This is possibly due to weathering which is particularly active for a wide fissure of the first generation. Fissures termed 'radial' (after Lebedev) or 'curvilinear' (after Hill) run inside the column. It is very difficult to decide between the two approaches, though Lebedev's hypothesis on the radial and concentric arrangement of thin fissures inside a joint column looks more justified if viewed in terms of stress field symmetry.

As observed by Hill (Figure 4.1(e)), pyroxene crystals in dolerite are located so that their long axes are perpendicular to small fissures. There is evidence of a similar arrangement of crystals in basalts.

Platy jointing formed by fractures of one system parallel to the intrusion contact succeeds columnar jointing in a zone of exocontact up to a few metres thick. Also encountered in this zone is blocky jointing formed by a pattern of short interecting fissures of chaotic arrangement. Blocky jointing of dolerites is similar to that of basalts. Open fissures and cavities are observed in the zone of blocky jointing at the contact of an intrusion with country rock.

Spheroidal jointing in intrusive dolerites can also be accounted for by

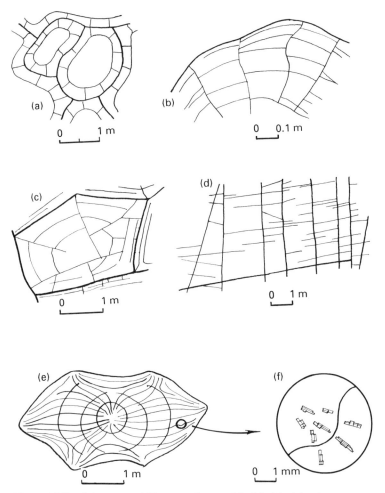

Figure 4.5 Small fractures within trap columns: (a), (b), (c), (e) column cross-section; (d) column longitudinal section; (f) arrangement of pyroxene crystals at fracture ((a), (b) after Lebedev, 1955; (e), (f) after Hill, 1965)

cavities in the intrusion zone. One can see blocks with shell-like jointing among dolerite blocks of irregular shape at the floor of a sill on the Angara.

4.4 Gabbro fractures

Large basic plutonic intrusions of differentiated gabbro and norite are accompanied, much like traps, by granitoid rocks (granophyre or granite) occurring at the top of basic rock bodies. Both gabbro and trap formations feature similar structural environments and mineralization. The major differences include the dimensions of intrusive bodies and the degree of their differentiation which depends upon rate of cooling and consequently upon the depth at which magma crystallization took place. It is due to this that the basic rocks of the two formations are characterized by different texture and fracturing.

The formation of differentiated gabbro and norite intrusions is represented by stratified bodies which occur in slightly dislocated strata of Archean and Proterozoic platforms. The stratified bodies range in thickness from a few tens to several thousands of metres. Large bodies are fairly common, as is not the case with the trap formation.

The dominant rock types are gabbro, norite and diabase. Pyroxenites, peridotites, dunites and anorthosites also occur. All these basic and ultrabasic rocks are products of deep differentiation of magma in an intrusion chamber. Cross-cutting acid rock bodies, associated with melting of country rocks, are sometimes encountered at the top of the intrusion.

Fracturing in the Pilguyarvi rock mass (Pechenga ore field, Kola Peninsula) is related to the formation of differentiated gabbro and norite intrusions (Mazanik and Makarov, 1970; Alekseyev, 1978). This differentiated basic–ultrabasic rock mass occurs among tuffaceous sedimentary rocks associated with graben-synclines and bounded by intraformational faults. Subparallel to the strata of country rocks, it is sheet-like in form (Figure 4.6). Its roof and floor dip at 45–55°. The rock mass is composed of serpentinized peridotites and serpentinites (average thickness 140 m), pyroxenites and gabbroids. The rock mass is ascribed to the Archean, ca 1890 Ma.

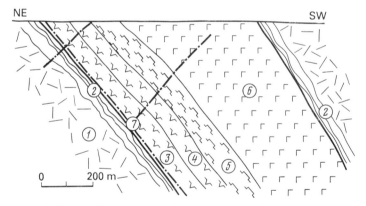

Figure 4.6 Schematic geological section, Pilguyarvi rock mass 1 – volcanic country rocks; 2 – hornfels; 3 – serpentinites; serpentinized peridotites; 4 – serpentinized perodotites; 5 – pyroxenites; 6 – gabbro; 7 – longitudinal and transverse tectonic faults (Alekseyev, 1978)

The set of original fractures displays systems of longitudinal fractures running along linear structures (system S after Cloos), transverse system Q perpendicular to the linear structures and fractures of system L which are, as a rule, parallel to the lower contact of the rock mass. The original fractures are healed by veined formations generically associated with basic–ultrabasic magma: gabbro–pegmatites and gabbro–diabase dykes.

There are contraction, tectonic and hypergene fractures. Contraction fractures have a uniform distribution. Late tectonic fractures were formed under plastic deformation. Release and weathering fractures in the hypergene zone rejuvenate the ancient system of fractures, the effect of hypergenesis being traced down to 30 m.

Of interest is the difference between original fractures in ultrabasic and basic rocks in the Tolstomysovsky trap intrusion (see Section 4.4). Here the late differentiates, in this case gabbro, differ from the early ultrabasic differentiates in the pattern of fractures (Table 4.1). The fracture systems in gabbro, peridotite and pyroxenite are identical. However, the gabbro fractures follow a more uniform pattern. Fracture spacing in gabbro exceeds that in ultrabasic rocks of smaller grain size. All this suggests that gabbro was the last to crystallize under the conditions of slow heat transfer between the intrusion and country rock. The earlier ultrabasic rocks feature high fracturing dispersion, thus pointing to high mobility of country rock. Noteworthy is the fact that the central portion of the rock mass composed of gabbro has solid blocks measuring 8–14 m^3. It is there that fracturing was latest and slow.

Table 4.1 Rock fracturing outside tectonic crush zones

| | | Fracture set parameters | | |
| | Fracture | | | average fracture |
Rock	set no.	azimuth	dip angle	spacing (cm)
Serpentinized	1	180–200	60–80	0.56
Peridotite	2	36–60	35–60	0.53
Pyroxenite	3	310	70	1.20
	1	215–230	30–60	0.61
Gabbro	2	40–55	50–55	0.40
	3	315–320	68–80	1.40

In Pilguyarvi the rock blocks are different in size. In peridotites and serpentinized pyroxenites the blocks range from 0.22 to 0.27 m^3, whereas in gabbro they are about 0.35 m^3.

In the fault zones the fractures are more closely spaced (by 1.5–2.0 times) and new fracture systems originate. The pattern of fractures in the ultrabasic rocks in the intrusion flexure zone is fairly complex. The arrangement of the major fracture systems changes according to the spatial position of the rocks. Fracture spacing tends to decrease. New fracture systems are observed at the contact in the flexure zone. Thus, the Pilguyarvi rock mass displays a regular increase in fracture pattern heterogeneities towards the contact and tectonic deformation zones.

Tectonic fracturing in gabbro is on the whole similar to the pattern of fractures in the ultrabasic rocks. Yet it is less pronounced. The differences between gabbro and the ultrabasic rocks in the crush zones are presumably inherited from the original fracture set. Otherwise, this fact can hardly be accounted for since the difference between the rocks as to their mechanical properties is minor.

Generally speaking, the fracture set in the basic–ultrabasic intrusion has much in common with the pattern of fractures in batholiths and trap sills.

Salient features of fracturing in igneous rocks

Fracture sets in effusive and plutonic intrusive bodies are essentially different. The former display polygonal, spheroidal and chaotic fracture patterns, both persistent and sub-persistent. Regular systems are very rare and fractures are mostly open. As for the latter, plutonic intrusive bodies, they feature regular fracture systems. Chaotic systems are observed in very small numbers. Original fracture systems are sub- or non-persistent. The fractures are filled with hydrothermal deposits. The pattern of fractures in hypabyssal bodies is varied: it resembles either that of extrusions or of plutonic rocks.

Tectonic and hypergene transformations of the fracture system are mostly inherited due to high rock strength. In igneous rocks disjunctive dislocations are more common than folded. The folds are gentle and complicated, as a rule, by faults. The salient features of tectonic fracturing in the igneous rocks are as follows: the rectilinear character of the zones of increased fracturing; a tendency for density within a fracture system; mylonitization and crushing of the rocks along the fractures; the presence of open fractures and hence high water permeability.

Observations on the successive generations of fractures in trap sills support the conclusion that fracture parameters regularly change with time.

The major differences between fracture systems in igneous rocks and those in sedimentary rocks may be summarized as follows: widespread occurrence of non- and sub-persistent fracture systems in plutonic intrusive rocks; a greater variety of fracture systems in igneous rocks, namely the presence of chaotic and regular systems with non-orthogonal sets; the presence of original microfracturing or splitting along blind joints.

4.5   Fractures in the Tolstomysovsky trap sill

Many of the features in this chapter may be exemplified by fractures in the Tolstomysovsky trap sill.

Geology

The Tolstomysovsky trap sill crops out in the middle course of the Angara. Coming out to the day surface is a thick trap intrusion of Triassic age overlain in places by sedimentary rocks. The geology of the area is schematically represented in Figure 4.7. The right and left banks as well as the watercourse area are composed of dolerites ranging in thickness from 60 to 100 m. The top of the sill is almost fully exposed by the river, the dolerites proper being practically unaffected by washout. On the left bank in the Tolsty Mys[*] (Thick Cape) hill, there is a swollen area of the intrusion

---

[*] Hence the name of the trap sill.

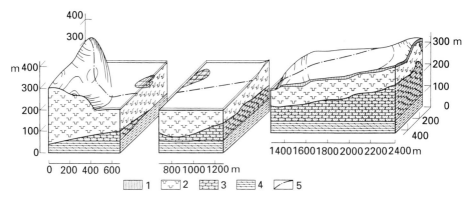

Figure 4.7 Tolstomysovsky trap sill: 1 – Quaternary deposits; 2 – dolerites; 3 – sandstones; 4 – siltstones; 5 – dam axis

250 m thick. The swell is in fact a hill whose relatively flat top has an elevation of 100–120 m above the river. There is a downwarp in its central portion which contains sedimentary rocks, while the northern part has an uplift with an elevation of 140 m above the water level in the river.

Two of the three slopes of the swell were freed from country rocks by erosion and are in fact steep slopes of the hill. The third steep slope faces the left bank and is almost fully overlain by sedimentary rocks which compose the saddle between the bank and the dolerite hill. Traps are seen to occur in all directions from the swell as a seam non-persistent in thickness which is complicated by some minor bulges and bends. In the river channel dolerite thickness decreases as we move from the left to the right bank and from the tail race to the head. At the left bank in the tail race the dolerites are 160 m thick. They pinch out at the right bank in the head race. On the right bank, much like in the river bed, the trap floor is uplifted south-eastward, reaching a maximum elevation (50 m) above the water line in the south-eastern part of the area. There it approaches the roof so that the interval between those is just a few metres.

The dolerite is a crystalline dark-grey or almost black rock composed mostly of monoclinic pyroxene and basic plagioclase. The crystal size ranges from fractions of a millimetre to a few millimetres, averaging 1 mm. The mineral crystals are characterized by a chaotic arrangement (massive texture) and occasionally give rise to accumulations, thus imparting a spotted appearance to the rocks. The pyroxene crystals fill the gaps between the plagioclase, the former being much larger (poikilo-ophitic texture) or equal (ophitic texture) in size. Apart from the above-mentioned rock-forming minerals, the rock comprises accessory minerals such as olivine, quartz, anorthoclase, titanomagnetite, marcasite. The dolerites are clearly differentiated. One can see the upper zone with granophyre material and the lower zone comprising large amounts of olivine. These zones are separated by a zone of dolerites of normal composition. Normal dolerites also occur along the contact between traps and sedimentary rocks. It is only in the immediate vicinity of the contact that the dolerites are represented by porphyritic and microporphyritic varieties.

Set of original fractures

The main stages of fracturing of the Tolstomysovsky traps are (i) original fracturing contemporaneous with cooling of the intrusion; and (ii) exogenic fracturing contemporaneous with formation of the present denudation surface.

Fracturing is described on the basis of 33 000 measurements of fracture parameters (dip azimuth, dip angle, spacing, width, length, wall roughness). A total of 14 000 fracture have been measured.

A notable feature of the original fractures is that they are filled with calcite and chlorite deposited from hydro-thermal solutions genetically associated with the trap magma. The fractures may be grouped with the original ones because they contain hydrothermal minerals.

The arrangement of the original fractures in the Tolstomysovsky sill follows on the whole the same pattern as do traps in general. The sheet-like portion of the intrusion displays mostly vertical and horizontal fractures that give rise to columnar and other types of jointing. Figure 4.8 shows systems of fractures with subvertical and subhorizontal dip angles. The differentiation of fractures in terms of dip can be observed throughout the sheet-like portion of the intrusion.

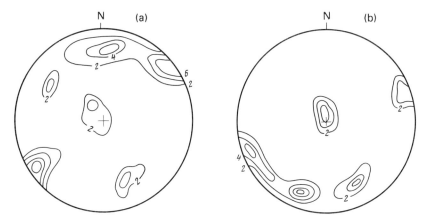

Figure 4.8 Circular diagrams of trap fracturing in the sheet-like portion of the Tolstomysovsky intrusion, stations 46(a) and 47(b). Point density isolines are expressed on a percentage basis

The differentiation of fractures into two categories, subvertical and subhorizontal, is most pronounced in the inner portion of the intrusion composed of normal ophitic and olivine troctolite dolerites. Fractures dipping at 30–50° are extremely rare (Figure 4.9). Fractures with dip angles ranging from 70 to 90° account for about 85% of the total number of

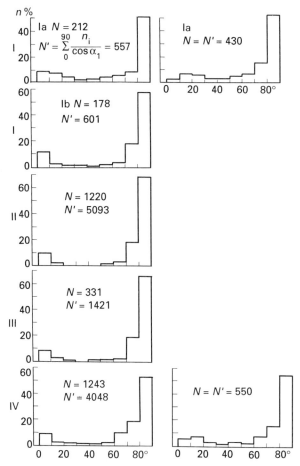

Figure 4.9 Distribution of fracture dip angles in the sheet-like portion of the Tolstomysovsky intrusion, core (left) and outcrops (right): Ia – upper near-contact zone, 0–10 m; Ib – same, 10–20 m from contact; II – inner zone; III – lower near-contact zone, 0–10 m on contact; IV – inner zone in the area of increased fracturing at the right bank; N – measured number of fractures found in core; N – corrected number of fractures – actual number of fractures in isometric volume.

fractures, while those with dip angles less than 30° make up approximately 12%. The variety of dip angles increases as the intrusion contacts are approached. For instance, some 10–20 m away from the upper contact, where granophyre dolerites are generally widespread, there are fractures dipping at various angles. The percentage of subvertical and subhorizontal fractures decreases somewhat. The fractures with dip angles ranging from 70 to 90° account there for 75% of the total number of fractures, a 10% decrease as compared with the inner portion of the intrusion. The dip angles are ever more varied directly at the contact (interval 0–10 m) in fine- and micrograined ophitic dolerites and dolerite–porpyrites. There, subvertical fractures dipping at 70–90° make up about 60%, a 25% decrease as compared with the inner portion of the intrusion. As the lower contact is approached, the dip angles also tend to vary, though over much narrower limits.

For precise quantitative assessment of variations in the fracture dip angles we have calculated their standard deviation from mathematical expectation. Dip angles $\beta = 0$ and $\beta = 90°$ were taken as the mathematical expectation for subhorizontal and subvertical fracture systems respectively. Depth-dependent variations of the standard deviations are shown in Figure 4.10. There is a distinct tendency for smaller s_β as we move away from the contact. The s_β variations are different in the case of subvertical and subhorizontal fractures and at the lower and upper contacts. In the latter case they can be approximated by exponential and logarithmic functions.

For subvertical fractures the relation is assessed by $\eta_{s/h} = 0.63$ and for subhorizontal fractures, by $\eta_{s/h} = 0.76$. As is evident from the diagrams, the upper contact has an effect on the arrangement of fractures down to within 2.5–3.0 m of the lower contact. For subhorizontal fractures such a relationship is traced to the middle of the intrusive sill or somewhat deeper.

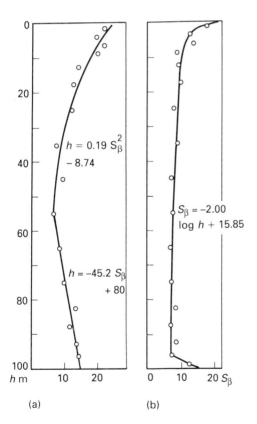

(a)   (b)

Figure 4.10 Variation, in the intrusive section, of root-mean-square deviation of dip angles of subhorizontal (a) and subvertical (b) fractures; distance from upper contact (h), sill thickness 100 m

The s_β variations both for subhorizontal and subvertical fractures are smaller at the lower contact than they are at the upper contact. For subvertical fractures these variations are so minor that they can hardly be approximated by any function. For subhorizontal fractures they are more clearly defined. The effect of the lower contact on the arrangement of subhorizontal fractures is traced 40–45 m away from the contact. The s_β shows a linear decrease as we move away from the contact. The relationship between s_β and h is assessed with the correlation factor $r = 0.84$ different from zero with reliability $P = 0.99$.

Summarizing the description of fracture dip angles in the sheet-like portion of the intrusion, the following conclusions may be drawn: (i) fractures are clearly differentiated into subvertical and subhorizontal; (ii) spread of dip angles assessed by standard deviation is a minimum in the inner portion of the intrusion and shows a linear increase in the direction of the lower contact; (iii) the upper contact has a much more pronounced effect on the arrangement of fractures than the lower contact; (iv) changes in the arrangement of subvertical fractures at the contacts are more drastic than they are in cases where subhorizontal fractures are involved; in the inner portion of the intrusion the arrangement of subvertical fractures is much more stable than that of subhorizontal fractures.

An increase in the variety of fracture dip angles in the sheet-like portion of the intrusion is observed not only when moving from the inner portion of the intrusion to the contacts, but also from the central portion of the intrusion field to its flanks. The intrusive sill becomes more complex in shape as the marginal areas of the intrusion are approached. Hence the arrangement of fractures also becomes somewhat more complex. Figure 4.9, IV, shows the distribution of fracture dip angles in this portion of the intrusion. Here, too, one can see mostly subvertical and subhorizontal fractures. Yet even at a considerable distance from the contact fractures of any direction are encountered.

The trend of subvertical fractures in the sheet-like portion of the intrusion may be grouped into two mutually perpendicular systems. The dominant trend of subvertical fractures is north-easterly and south-westerly. The same trend has been observed while studying fractures in traps at the foundation of the Vilyui and Bratsk Hydros. They can be treated as a major background against which there are anomalies associated with some specific features of the stress field. These anomalies are fairly common in the Tolstomysovsky sill. Trending persistently, the systems of vertical fractures are traceable in some portions of the intrusion.

One of those has subvertical fractures grouped into two systems, one system with a strike azimuth of 320–340° and the other with a strike azimuth of some 60–80°. It is interesting to note that the intrusive sill is persistent in dip and strike in this area with no bulges or bends (Figure 4.7). One of the vertical fracture systems is arranged down the dip and the other along the strike of the intrusive sill. Following Cloos, these systems may be referred to as longitudinal (S) and transverse (Q). There is reason to believe that the persistent trend of the fractures results from that of the intrusive sill due to which there was a relatively homogeneous anisotropic stress field during original fracturing throughout the area in question.

In the intrusive dome, fractures follow a fan-like pattern, which is typical of traps (Figure 4.11). At the contact, fracture sets are either chaotic (Figure 4.12), giving rise to finely blocky jointing or regular (Figure 4.13), giving rise to thin platy jointing. At a distance of several metres from the contact the fracture set takes the regular or polygonal form (Table 1.1). In the central part of the dome the fractures are arranged chaotically.

In the zone of columnar jointing there are two systems of steep fractures and one system of flat fractures. There is a tendency for the fracture planes to intersect at right-angles, giving almost rectangular joint blocks, as in the sheet-like portion of the intrusion. The fracture systems bounding the joint

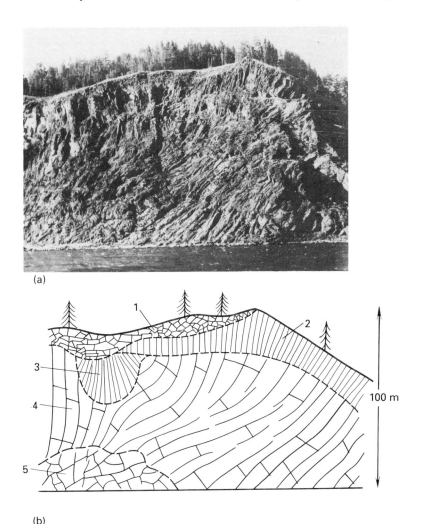

(a)

(b)

Figure 4.11 General view of inner portion of the dome-shaped swell of Tolstomysovsky sill: (a) photograph; (b) diagram; 1 – blocky jointing at contact; 2 – finely columnar; 3 – fan-like; 4 – coarsely columnar; 5 – blocky, central portion of dome-shaped swell

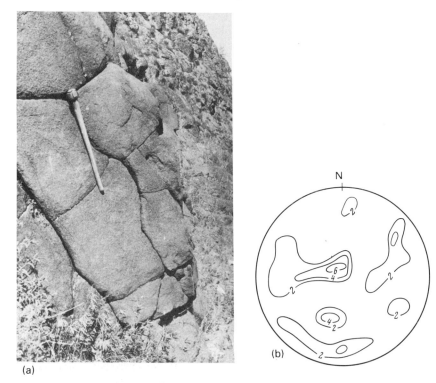

Figure 4.12 Chaotic system of fractures in central portion of the dome-shaped swell:
(a) photograph; (b) diagram, point density isolines expressed on a percentage basis

blocks are characterized by some particular arrangement relative to the roof of the intrusion. In the immediate vicinity of the contact the flat fractures are parallel to the roof; moving inside the intrusion, their dip changes and they assume an intermediate position between the horizontal fractures and those running parallel to the contact. One of the steep fracture systems, conventionally termed 'longitudinal', is parallel to the strike of the contact and dips in the opposite direction. This system is in between the vertical and the plane perpendicular to the contact. The other system of steep fractures is parallel to the dip of the contact, the fractures being conventionally termed transverse. Their dip angles are close to vertical.

The pattern of fractures in the dome-shaped portion of the intrusion is simplified as described above. Actually, it is much more complex. Apart from the major fracture systems, there are additional systems.

Density of trap fractures, much like their arrangement, varies from one portion of the intrusion to another. These variations follow a pattern common to traps. The spacing between neighbouring flat and subhorizontal fractures varies, as does their arrangement, when moving from the inner portions of the intrusive sill to the periphery (Figure 4.14). In the inner portion of the intrusive, the fractures are widely spaced

(a)

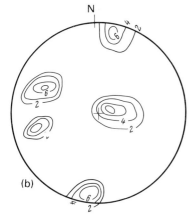

(b)

Figure 4.13 Regular system of fractures at contact in the dome-shaped swell: (a) photograph; (b) diagram, point density isolines expressed on a percentage basis

(0.5–0.6 m) apart, whereas at the contacts the spacing decreases by nearly one-half. This process is parallel to the change in the rock petrography. Therefore, dolerites of different petrographic composition feature differently spaced fractures (Table 4.2).

At the lower contact of the intrusion, variations in the density of horizontal fractures follow a linear law in the endocontact zone whose thickness makes up 20% of that of the sill. The relationship is assessed with the correlation factor $r = 0.72$. The effect of the upper contact on the density of horizontal fractures is traced down to 80 m, i.e. for 80% of the sill thickness.

Variations in the density of subvertical fractures follow a different pattern. Firstly, it should be noted that fracture spacing is fairly small throughout the intrusion. The core samples display all minor fractures of the second and later generations in trap columns (see Figure 4.5).

Subvertical fractures, much like subhorizontal ones, are fairly dense at the upper and lower contacts. They become spaced more widely as we

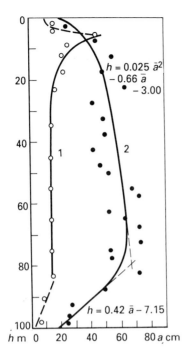

$$h = 0.025\,\bar{a}^2 - 0.66\,\bar{a} - 3.00$$

$$h = 0.42\,\bar{a} - 7.15$$

Figure 4.14 Variations in average fracture spacing in the sheet-like portion of the intrusion: h – distance from upper contact, $h = 100$ m – lower contact position; 1 – vertical fractures, 2 – horizontal fractures

Table 4.2 Spacing of flat (dip angle 0–20°) fractures in dolerites of different petrography

Petrographic varieties	Number of fractures measured	Average fracture spacing
Ophitic I	72	0.28
Granophyric	341	0.53
Ophitic II	520	0.49
Troctolitic	611	0.58
Ophitic III	221	0.30
Total	1775	–

move away from the lower contact for 15–20 m. At this point the effect of the lower contact terminates. In the inner portion of the intrusion, for a depth range of 30–80 m from the upper contact, vertical fractures feature minor variations in the density. At the upper contact the relationship between fracture spacing and distance to the contact is fairly complex. The fractures are closely spaced in the first five metres, which may be attributed both to the effect of the contact and to weathering since the denudation surface coincides with the contact. In this interval fracture spacing increases with depth to a maximum of 44 cm at a depth of 5–7 m. Below this level, fracture spacing decreases as a hyperbolic function, which reflects the characteristic features of original fracturing. Variations in the dispersion of dip angles of horizontal fractures and in the fracture spacing with distance from the contacts, and the difference between the patterns of fractures in the dome and in the sheet-like portion of the intrusion, and some other fracturing parameters, are accounted for by the evolution of stresses in different parts of the intrusion.

Based on the morphology of the original fracture set in the Tolstomysovsky intrusion, the following scheme of original fracturing can be proposed.

Fracturing starts at the two contacts following crystallization of porphyritic microdolerites in the period of kinematic and dynamic activity of the country rock and magma. At the lower contact fractures grow more slowly than at the upper contact and at some point they may even cease to grow. Fractures growing from the bottom extend to the lower 15–20% of the sill thickness and those growing from the top pierce through the whole intrusion until they meet the fractures which grow from the bottom. It was thought that horizontal fractures in sills generally lag behind vertical fractures in growth. These investigations provide support for this view. Indeed, horizontal fractures, as opposed to vertical, are formed in less homogeneous environments and therefore feature a larger dispersion of their parameters.

The kinematic and dynamic activity of the upper and lower contacts is essentially different. The upper contact was apparently more mobile and it has a more pronounced effect on the fracturing process.

Hypergene transformations of fracture sets

The last stage in fracturing of the trap intrusion under consideration, and generally of traps in the region, involves their exposure to the ground surface and hence removal of gravity load, weathering, and the effect of slope stability. In the valley sides and in its floor, deformations show themselves differently and so it is advisable to consider them separately.

The traps composing the valley slopes feature release fractures, landslides, i.e. all known deformations peculiar to rock slopes. Stress release is favoured by some specific features of fracturing. Traps display, as noted above, original fractures close to vertical and horizontal. In the process of stress release these expand to give rise to cavities parallel to the slope.

A study of the stress release zone has been made in adit number 1 driven on the left bank at a point where the dome-shaped portion of the intrusion grades into its sheet-like portion. Identical fracture systems can be seen whose location corresponds to that of original fractures. No additional fracture systems are formed. Fracture systems are most clearly defined at the adit mouth. Further along the adit, the fracture patterns tend to vary, which agrees well with ideas on the original fracturing of the dome-shaped swell.

Fracture spacing varies little in all the system (Figure 4.15). Not so with fracture width or aperture which varies over fairly wide limits. Fracture width variations are most pronounced in a system more or less parallel to the valley slope (Figure 4.16). The average width of fractures regularly decreases along the first 5 m of adit depth and then remains approximately constant. At the adit mouth fractures vary in width. A special graph serves

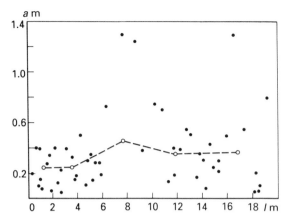

Figure 4.15 Variations in spacing a between vertical fractures more or less parallel to the slope along the adit length, l

to illustrate the dispersion of points, showing the dispersion of fracture width regularly decreasing inside the slope. Therefore, stress release is due to the fact that some fractures progressively expand, whereas others do not at all. In this case original fractures presumably exceed in number those required for stress release. Hence no new fractures originate, and not all the existing fractures expand.

One can also see that fracture filling changes due to weathering as the surface is approached. At depth the fractures are filled in solely with calcite and chlorite, which indicates that there are only original fractures. At the adit mouth the fractures are filled in with loam, iron oxides, water and air. Fractures with evidence of weathering are shown by triangles (see Figure 4.16).

In some instances, there also occurs so-called detachment, giving rise to erratic blocks parallel to the slope and separated from the latter with troughs up to 10 m wide and 15 m deep.

For instance in the Badarminsky contraction, the process of slope detachment is favoured not only by some specific features of trap fractures, but also by those of the geology of the area. The trap intrusions are fairly thin and fully cut by the river. In the river bed, trap-hosting Carboniferous terrigenous rocks are exposed due to erosion. These rocks are fairly soft. Under these conditions, when the lower portion of the valley slope is composed of soft rocks and the upper one, of trap seams, the slopes are undercut and the traps are shifted towards the river bed with intrusion discontinuity along the vertical original fractures. As a result, the two slopes of the Angara valley in this area are composed of slumps some 2–3 km away from the river and the watercourse is half as wide as it is in other contractions.

These slumps were named the Angara-type block landslides by Sokolov (1961), who found them in sedimentary rocks in the Angara valley. Detachment and slumping of the river-cut traps in the Badarminsky contraction is not a rare phenomenon. These processes may also be observed elsewhere in the area.

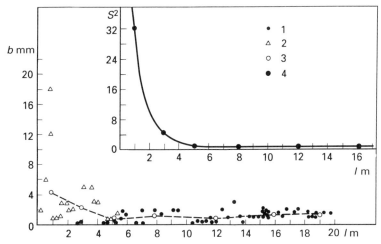

Figure 4.16 Variations, along the adit length, in fracture width, b, and its dispersion in a set more or less parallel to the Angara valley slope: 1 – fractures devoid of weathering traces; 2 – those with weathering traces; 3 – average fracture width; 4 – width dispersion

The aforesaid disturbances are associated with the valley slopes. However, there are also deformations in the floor of the valley resulting from stress release due to erosion. They take place, as a rule, under the river channel and hence are not visible. Studies of these deformations are of practical importance in the construction of hydraulic projects. There are two types of release deformations, those of the first type being described by Tizdel (1962) who attributed them to a fracture in a rock stratum brought about by elastic expansion of underlying rocks. Deformations of the other type occur to a depth of 30 m. The Tolstomysovsky sill serves to illustrate these two types of deformation.

The fact that rock permeability increases as the surface is approached points to the presence of a stress release zone with open fractures. In the zone of open fractures, their arrangement is not different from the arrangement in the portions of the intrusion unaffected by stress release. The density of subhorizontal fractures, namely fractures parallel to the erosional valley surface, changes slightly, whereas the density of subvertical fractures is not apparently increased. There are also changes in fracture filling. Instead of calcite and other hydrothermal minerals which infill fractures outside the stress release zone and the zone of weathering, fractures at the erosional surface are filled in with loam, sand, iron oxides, and water.

Fractures 10–50 m long were traced in the river bed. The width of NW-trending fractures ranged from 1 to 5 cm and that of NE-trending reached 70 cm. Fractures were open to a depth of 0.5–1.5 m, mostly due to weathering and evacuation of the finely platy dolerite resulting from original fracturing along large original fractures of the first generation.

After studying the pattern of fractures in the Tolstomysovsky intrusion, it can be concluded that the pattern currently observed was formed during original fracturing and is somewhat different close to the surface due to due to stress release. The salient features of trap fractures are accounted for by the course of original fracturing and exogenic deformations.

Indirect evidence of sill fracturing

Direct methods for measuring fracture parameters provide comprehensive characteristics of fracture sets. The fracturing models based on such measurements are used in solving diverse problems of engineering and structural geology. However, poor exposure and lack of exploration openings prevent measurements from being taken as required. That is why extensive use is made of indirect methods for determining fracture parameters, such as intake of water by a borehole, determination of elastic wave velocity, observations on the direction and velocity of subsurface water flows and hydrochemical observations.

These indirect methods were used in studying the Tolstomysovsky sill. As a result it became possible to identify and delineate zones with increased fracture opening.

A zone of increased dolerite fracturing was identified under the Angara bed through injections followed by seismic profiling. It features lower velocities of elastic waves and higher specific water absorption (10–100 l/m), average water absorption for traps being about 0.001 l/m. Increased fracturing is due to opening of closely spaced subvertical fractures to a depth of 30–50 m. Upstream the zone of increased fracturing widens due to a decrease in sill thickness. Downstream as the latter grows (Figure 4.7), the zone of increased fracturing pinches out. Increased fracturing is associated there with a bend of the sill where its thickness decreases. It results from a fracture in the sill following stress release due to erosion. In the release deformation zone, there are lenticular bodies of finely crushed dolerites 3–6 m long and 40–70 cm wide. Such areas are characterized by lower velocities of longitudinal waves and hence can be detected through seismic profiling.

Apart from identification of individual fracture zones, the indirect methods are instrumental in assessing changes in the pattern of fractures throughout a rock mass. Analysing the results of borehole injections makes it possible to determine the exposed width of the zone of increased water permeability at the intrusion contact. Plotted on the ordinate (Figure 4.17) is distance from the contact into the country rocks (upper portion) and into traps (lower portion). The country rocks are represented by sandstones and siltstones of the Tunguska coal-bearing series. Plotted on the abscissa is water permeability (log K_p). Permeability coefficients are based on individual injection into and pumpings out of cored holes.

As is evident from the diagram, water permeability is a maximum at the contact, gradually decreasing on passing into the intrusive body and the sedimentary rock mass. It tends to reach certain values typical of unaltered sedimentary and igneous rocks. In the near-contact portion water permeability has a geometric average (K_p) of 0.7 m/day. It shows a decrease in the direction of sedimentary rocks within a distance of 20 m ($K_p = 0.1$ m/day) and within an additional distance of 20 m to $K_p = 0.05$ m/day. Water permeability also decreases in the direction of

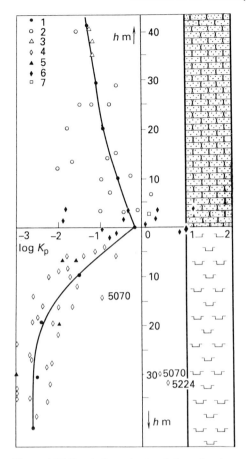

Figure 4.17 Depth-dependent variation of water permeability at a trap–sedimentary rock contact: 1 –average K_p in different depth ranges; individual results; 2 – in sedimentary rocks close to lower contact of intrusion; 3 – close to upper contact; 4 – in traps close to lower contact; 5 – same close to upper contact of intrusion; 6 – in depth ranges covering sedimentary rocks and traps – lower contact; 7 – same upper contact

igneous rocks to $K_p \approx 0.002$ m/day within a distance of 15–20 m, after which it remains constant.

The relationship between average water permeability and distance from the contact may be approximated by the following equations:

$$\log K_p = -1.3 \ln (h+30) + 4.2 \text{ for sedimentary rocks}$$
$$\text{Log } K_p = -0.88 \ln (h+0.52) \text{ for traps}$$

Judging by the correlation degree, water permeability of the country rocks, although decreasing away from the traps, depends little upon the distance from the contact. At the same time higher water permeability at the contact is typical of traps. It may be inferred that there is a zone of increased water permeability at the contact of the trap intrusion. Its width can be determined at any specified water permeability. For example, a zone with permeability coefficient $K_p \geqslant 0.1$ m/day is about 20 m wide and the one with a permeability coefficient of the order of 1.0 m/day is about 4 m wide.

Figure 4.18 illustrates variation in log K_p dispersion. In sedimentary rocks dispersion decreases as a linear function to 25–30 m from the contact inside the rock mass. It remains almost constant 10–40 m from the contact in traps.

The thickness of a zone with varying water permeability and the nature of these variations in the zone of weathering and stress release can be solved statistically through the analysis of experimental data. Results of the permeability studies used were made in the same rock and under similar weathering conditions, that is, under the river bed where dolerites of the inner portion of the sheet-like body are weathered. Water permeability of rocks unaffected by weathering under the river bed is not constant, but varies so that in some areas water permeability of unweathered rocks exceeds that of weathered rocks or vice versa. To assess the effect of weathering on water permeability under such heterogeneous conditions, water permeability variation due to weathering should be considered rather than a permeability factor (q or K) as in the case of the near-contact zone. Water permeability variation may be assessed with the relation

$$K/K_{av}$$

where

    K is the permeability coefficient obtained from individual pumpings out of a borehole;
    K_{av} is the average value of permeability coefficient for unweathered rocks based on data obtained from the same hole.

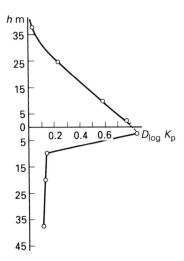

Figure 4.18 Water permeability dispersion at the contact

Similarly,

q/q_{av}

where

q is specific water absorption in a particular depth interval, l/m,
q_{av} is average specific water absorption for unweathered rocks,

may be used.

Depth-dependent variation of water permeability from the erosional surface is shown in Figure 4.19. The ordinate is depth m and the abscissa is the above-mentioned water permeability variation factor plotted on a logarithmic scale since in this form it features distribution closest to normal. It is evident that most points cluster around

$\log (q/q_{av}) = 0$

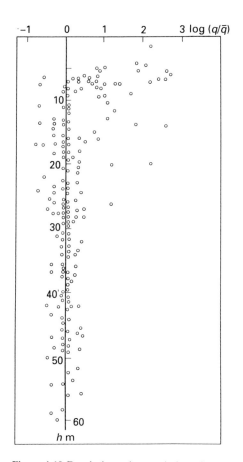

Figure 4.19 Depth-dependent variation of water permeability $\log (q/q_{av})$ under the Angara river bed

i.e. generally

$$q \simeq q_{av}$$

Hence it is clear that stress release and weathering have a minor effect on variation of water permeability in the area under study. Yet certain values greatly deviate from

$$\log(q/q_{av}) = 0$$

in the upper portion of the diagram. Water absorption may increase by nearly 1000 times due to fracture opening or decrease by nearly 100 times due to fracture grouting or sealing (colmatage). On the whole the average valueof $\log(q/q_{av})$ in the upper portion of the diagram exceeds zero, which indicates that water permeability increases due to weathering. However, the spread of the points is so large that there is only a minor relationship between water permeability and depth. The correlation coefficient calculated on the basis of 190 tests is 0.33.

The relationship may be approximated by the equation

$$\log(q/q_{av}) = - a \ln(h+b) + c$$

where
  h is distance from the erosional surface;
  a, b, c are certain positive numbers.

The spread of the points is different depthwise. Depth-dependent variation of distribution is shown in Figure 4.20. The abscissa is dispersion of individual values $\log(q/q_{av})$ calculated for some depth intervals and the ordinate (h) is depth in metres. The points are arranged so that they form a broken line, a portion with the equation

$$s^2 = 1{-}0.03\ h$$

being located in a depth interval of 0–30 m.
  A portion of the line with the equation

$$s^2 = 0.05$$

is located in a depth interval of 30–60 m. Hence the effect of stress release and weathering does not extend below 30 m.

The foregoing zones of increased water permeability at the contact and in a zone of weathering are fairly common. Near-contact transformations

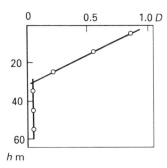

Figure 4.20 Variation in dispersion D of water permeability $\log (q/q_{av})$ under the Angara river bed

of fracturing can be observed in all igneous rocks, while opening and colmatage of fractures in the hypergene zone are presumably typical of all rocks without exception. There is another interesting feature of the Tolstomysovsky sill, apparently observed for the first time for igneous rocks: water permeability and hence fracture opening are seen to depend upon sill thickness (Figure 4.21).

As is evident from the distribution of points, there is a relationship between the random value log q_{av} and thickness of the intrusion, which is characterized by the correlation coefficient $r = 0.80$ different from zero with reliability $P = 0.99$. Higher water permeability with smaller thickness of the intrusion is evident from the distribution of perched water on the traps. Perched water accumulates wherever the intrusion is thick, say, on the top of the dome-shaped swell where it is found all the year round. The dome-shaped swell is an isolated hill (Figure 4.22) where infiltration is the only source of perched water recharge. Assuming this to be equal to annual precipitation (ca 360 mm), one may calculate that a permeability coefficient $K = 0.01$ is sufficient to ensure a perennial supply of perched water. Actually recharge is some minor portion of the precipitation level because the ground above the perched water is mostly frozen and runoff is favoured by the relief features. Consequently, the average K_p in the swell is less than 0.001 m/day. One may obtain

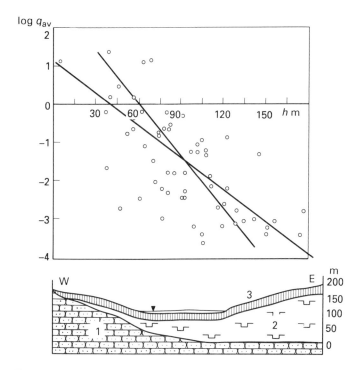

Figure 4.21 Schematic geological section illustrating variations in trap intrusion thickness and a graph of trap water permeability versus thickness: 1 – sedimentary rocks hosting the intrusion; 2 – trap intrusion; 3 – stress release and weathering zone

$K_p = 1 \times 10^{-6}$ m/day

in this portion of the intrusion through extrapolation (see Figure 4.21).

The presence of an isolated zone of increased water permeability which pierces through the intrusion (Figure 4.22) is evident from the chemistry of the groundwater. Traps have essentially ultrafresh water and it is only in the zone of deep exogenic deformation that trap groundwater has a mineral content of up to 2 g/l and a composition close to that of water entrapped in sedimentary rocks that underlie the traps.

Thus, the indirect investigations of fracturing agree well with the results of direct measurements of fracture parameters, i.e. the sill displays several zones with different fracture patterns.

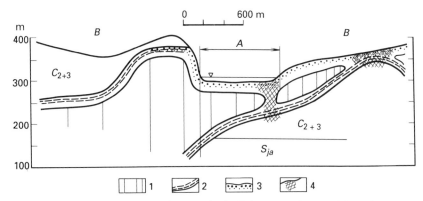

Figure 4.22 Zones in a system of fractures in the Tolstomysovsky trap sill: 1 – inner portion of the sill unaffected by near-contact and hypergene transformations of the fracture system; 2 – endocontact of intrusion with zones featuring wider or more closely spaced fractures; 3 – hypergenesis zone characterized by widening or partial filling of fractures; 4 – zones of deep fault-type deformation of traps due to rock mass stress release; A – river bed; B – bank; C_{2+3} and S_{ja} are sediments into which the sill is intruded

Summary

The set of fractures in the Tolstomysovsky trap sill has a structural pattern typical of trap bodies. It was formed in two stages. The first stage involved relatively uniform fracturing of the rock mass due to contraction. The parameters of this process and those of the fracture set regularly changed as a function of the form and thickness of the intrusion and the rate of cooling. The second stage was characterized by inherited evolution of the fracture set. The longest and widest fractures expanded. In areas where the intrusive sill pinches out (the fractures were closely spaced there due to the proximity to the contacts), there are fault-type deformations. These manifest as increased opening of fractures in some zones rectilinear both in section and in plan.

Part II

Methods for studying fracture sets to assess rock mechanics and macrostructure

Chapter 5
Field measurements of fracture parameters

5.1 Objectives and methods

Sets of fractures in rock masses are studied, firstly, to assess mechanical properties of rocks for predicting their stability in underground openings and slopes of open-pit mines, forecasting mine inflows, deformation of foundations and solving some other hydrogeomechanical problems. In this case the investigations are aimed at constructing a geometrical model of fracture sets, its parameters being defined by specific engineering problems.

Secondly, fracture studies are conducted to solve a variety of geological-structural problems, such as identification of faults and assessment of their parameters, and reconstruction of stress field patterns.

Geometrical models of fracture sets can be developed as a result of geological surveys at different scales. Small-scale models are based on aerial photographs (remote sensing), whereas large-scale models are constructed following special ground surveys, sometimes using photo-theodolite techniques.

There are some essential requirements to be imposed on a model of fracture sets. Any geometrical model is based on such parameters of fractures as dip azimuth (α) and dip angle (β). Use is also made of fracture width (b), fracture length (l) and fracture spacing (a) in a system. In addition to these principal parameters, some other parameters may sometimes be used as required, for example wall roughness, composition, compressibility and water permeability of fracture fillers.

The geometrical parameters of fracture sets are determined separately for each outcrop and for each particular area with homogeneous fracturing. Observational data obtained from different areas provide the basis for maps and sections which serve to assess variation of fracture patterns in a rock mass and the macrostructure of fracture sets. Therefore, a fracture set model is always consistent with the arrangement of large geological structures to which fractures are genetically related. This provides a better insight into these large structures, enabling one to assess the location of tectonic faults, fold axes, and intrusion contacts.

A fundamental problem in studying fracture sets is the construction of a quantitative model based on direct measurements of fracture parameters in a rock mass. Construction of a hypothetical set of fractures based on information on folds, faults and other structures is referred to as an inverse problem.

In the present state-of-the-art, solution of the inverse problem lacks accuracy due to considerable random variation of fracture sets. It is suitable only for the general layout of a construction project or may be used in planning geological exploration.

5.2 Large-scale geological survey of fracture sets

Large-scale geological survey of fracture sets involves direct measurements of fracture parameters in natural or artificial exposures.

Listed below are the principles of studying fracture patterns through large-scale engineering geological surveys.

1. Quantitative studies of fracturing are essential since it is necessary to digitize engineering–geological and hydrogeological forecasts.
2. A number of fracture parameters have to be determined due to the variety of fracture patterns. It is impossible to deal with individual generalized parameters.
3. Since large fractures have a major role to play in the processes under study, the following rule holds in accordance with the principle of inheritance, namely, the larger the fracture, the more comprehensively it is to be studied. Microfissures in a rock sample are not generally studied. In studying macrofractures, one employs techniques of engineering–geological surveys and mathematical statistics. Megafractures and tectonic faults are studied individually. Their assessment may involve special excavations or drilling holes.
4. The genetic analysis of fracturing is essential at any stage of investigation inasmuch as it provides the basis for extrapolation and interpolation of the results.

The number of fractures to be studied in some particular area of a rock mass with homogeneous fracturing is determined by the immediate objective and the dispersion of their parameters. It varies over wide limits from 100 to 1000, being a minimum in areas with simple fracture patterns (three sets of one genetic type) and reaching a maximum in zones of crush and weathering.

Studies of fracturing commence by assessing the requisite number of survey stations and individual measurements for each particular parameter. Key outcrops to be studied are outlined. Special emphasis is placed on rock lithology, tectonics and geological history.

The survey stations should be located so as to enable adequate description of all the geologico-structural areas and the rocks occurring therein, as well as zones with different degrees of exogenic transformations.

Field measurements of fracture parameters should be taken by two or three operators, one of them collecting information for subsequent processing on a computer. It is necessary to measure all the fractures in the outcrop, their number ranging from 100 to 400 provided 10–40 fractures are measured in each set. The minimum number of measurements is taken when there are 3–4 sets with a small spread of parameters.

The operations to be performed on the outcrop include its description, measurements of fracture parameters and graphic representation.

The measurements are taken as follows. First, the arrangement of fractures is determined using a surveyor's compass. For higher measuring accuracy one may also use an aluminium plate measuring 10×15 cm as a base for the compass. The plate is inserted into a fracture or put on its surface as an approximating plane. Then a folding rule (steel rule) is used in measuring other parameters.

Fracture spacing is measured at right-angles to the fracture plane, and fracture width or aperture may be measured using a fitter's metal rule and a magnifying glass. Accurate measurements can be obtained from a fracture photograph with a negative enlarged and no photographic printing.

There is a special instrument developed by S. E. Mogilevskaya (Anon., 1980) for use as a profilograph to study fracture wall roughness. It comprises a large number of needles interposed between two rubber-coated strips. The needles can be released and secured by adjusting the clearance between the strips. To obtain a profile, the needles are set at right-angles to the fracture wall and lowered onto the rough surface. A rough surface profile is reproduced on the paper. The instrument is suitable for use both on horizontal and vertical fractures. The length of the profile thus obtained is about 1 m. Several parallel profiles can be obtained along one surface as is done by a matrix method. The profile reproduction accuracy is 0.1 ± 0.05 mm depending on needle diameter.

The methods described here are widely used, and several automatic profilographs are described in the literature (Fecker and Rengers, 1971; Hoek and Bray, 1974; Brown, 1981).

5.3 Accuracy of fracture measurements in field investigations

Fracture parameters measured on outcrops and openings are used in calculating the modulus of deformation, permeability coefficient and in solving other problems. Thousands of measurements have been made to determine measuring errors and some other errors, say those arising from inadequate assessment of fracture length by the geologist or from interpolation of observational data. The field measurements of fracture parameters may be affected by three types of error, namely metrological, methodological and an error of analogy. A measuring metrological error is due to instrument inaccuracy or some other factors not related to the geological structure of the fracture system involved. By contrast, the other errors arise when the geologist schematizes the actual geological structure of fracture systems. There are schematizing errors for an individual fracture and for a combination of fractures. The first may be conventionally called the 'methodological error'. When measuring fracture arrangement, this error arises when the rough surface of the fracture is approximated by a plane. When measuring fracture width, this error is associated with schematizing of the space between the two fracture walls as a distance between parallel planes. When measuring fracture spacing, the methodological error arises from the fact that a neighbouring fracture is presumed to be parallel to the one in question. When measuring fracture length, the methodological error is due to the fact that a curvilinear fracture is treated as a straight or broken line. The error of analogy is based, as the name implies, on the assumption of analogy between an object to be measured and other objects of similar origin and structure.

The number of measurements taken to assess different types of errors is tabulated in Table 5.1.

Table 5.1 Number of measurements

| | *Fracture parameters* | | | |
Type of error	*dip angle and azimuth*	*width*	*spacing*	*length*
Metrological	800	111	–	–
Metrological and methodological	35	750	136	10
Metrological, methodological and error of analogy	250	55	161	101
Total	2400			

Fracture arrangement (orientation)

Checking of the measuring accuracy of fracture arrangement with a surveyor's compass was done under laboratory conditions. An error arising from the selection of a measuring point on the fracture surface was not allowed for.

For laboratory studies, an aluminium plate used in the field and modelling the fracture surface was rigidly secured to a theodolite tube. By changing the position of the plate, two geologists measured dip angles and azimuths using surveyor's mirror compasses independently. At the same time the dip angle and azimuth of the plate were determined to an accuracy of 1'. A total of 800 measurements of dip angles and azimuths were made. Based on these data, standard measuring errors were calculated. They proved fairly small both for dip angles and dip azimuths.

It is believed that a lower dip angle affects fracture arrangement measurements. However, this is not so. Mathematically, it should be remembered that fracture arrangement is a two-dimensional value which is statistically characterized by two-dimensional distribution. If errors in measuring dip angles and azimuths are considered separately, we may erroneously conclude that the measuring accuracy of fracture arrangement at low dip angles is fairly low.

Mathematically, it can be shown by introducing the following notions that the measuring accuracy of fracture arrangement at low angles does not depend on the azimuth measuring error:

1. Orientation of the plane is its spatial position relative to the coordinate axes X, Y, Z (Figure 5.1) with X directed towards the north; Y towards the east; and Z into the zenith. The position of the plane, as is known

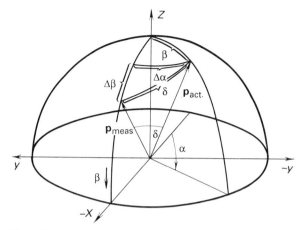

Figure 5.1 Measuring error (δ) of fracture orientation and its components ($\Delta\alpha$, $\Delta\beta$) on the coordinate sphere

from analytical geometry, can be given by vector **p** perpendicular to the plane. The vector is determined if we know: (i) angle α from the origin of coordinates to the projection of the vector on the horizontal plane X, Y; (ii) angle β in the vertical plane **p**Z between the vertical Z and vector **p**. In geology, α is referred to as the dip azimuth and β as the dip angle.

2. An error in measuring the spatial position of the plane is angle δ between the actual position **p**$_{act}$ and measured position **p**$_{meas}$. To determine this error, one should draw the plane through the vectors **p**$_{act}$ and **p**$_{meas}$ and measure the angle δ in this plane. The aforesaid angle δ is obviously an error in measuring the sought value in this particular problem. The same is true of problems of the tensor permeability theory.

From the standpoint of analytical geometry the fact that the δ error is independent of the $\Delta\alpha$ error at low angles β can be proved in the following way by showing that the angle δ tends to become infinitesimal should the errors in measuring dip azimuth ($\Delta\alpha$) and dip angle ($\Delta\beta$) decrease. For this it is enough to show that $\cos\delta \rightarrow 1$ when $\Delta\alpha \rightarrow 0$ and $\Delta\beta \rightarrow 0$, and that $\cos\delta \rightarrow 1$ on condition that $\Delta\beta \rightarrow 0$ and $\beta \rightarrow 1$ whatever the mode of $\Delta\alpha$ variation.

If this is proved, it can be asserted that with the dip angle falling to zero, an error in measuring orientation does not depend on the azimuth measuring error. Consequently, it is proved that with the dip angle falling to zero and the error in measuring dip azimuth increasing, the error in measuring the spatial position of the plane does not increase.

The cosine of the angle between two vectors of unit length (x_1, y_1, z_1) and (x_2, y_2, z_2) in the rectangular coordinate system is as follows:

$$\cos\delta = x_1 x_2 + y_1 y_2 + z_1 z_2$$

With the axes positioned as shown in Figure 5.1, the rectangular coordinates are expressed through the angles α and β as follows:

$$x_i = \sin\beta_i \cos\alpha_i; \ y_1 = -\sin\alpha_i \sin\beta_i \atop z_i = \cos\beta_i \Bigg\} \ i = 1,2$$

Therefore,

$$\cos\delta = \sin\beta_1 \cos\alpha_1 \sin\beta_2 \cos\alpha_2 + \sin\beta_1 \sin\alpha_1 \sin\beta_2 \sin\alpha_2 + \cos\beta_1 \cos\beta_2$$

where

$\beta_2 = \beta_1 + \Delta\beta; \ \alpha_2 = \alpha_1 + \Delta\alpha$
β_1 and α_1 are coordinates of the first vector in Figure 5.1;
β_2 and α_2 are coordinates of the second vector in vector in Figure 5.1.

After transformations

$$\cos\delta = 2\sin\beta_1 \sin(\beta_1 + \Delta\beta) \cos^2 (\Delta\alpha_1/2) + \cos(2\beta_1 + \Delta\beta)$$

Considering the equality provided $\Delta\beta \to 0$ and $\Delta\alpha \to 0$,

$$\cos\delta \to 2\sin^2 \beta_1 + \cos2\beta_1 = 2\sin^2 \beta_1 + \cos^2 \beta_1 - \sin^2 \beta_1 = 1$$

Hence, with errors in measuring dip angle and azimuth decreasing to zero, the error in measuring the spatial position of the plane also tends to zero.

Considering the same equality provided $\Delta\beta \to 0$ and $\beta \to 0$, then $\cos\delta \to 0 + \cos 0 = 1$.

Thus, if the plane dip angle tends to zero, the error in measuring the position of the plane ceases to depend upon $\Delta\alpha$. To assess the orientation measuring error, the results of the laboratory experiment with a theodolite are plotted on a circular diagram of fracturing (Figure 5.2) as bivariate error distributions. Located in the different parts of the diagram, they characterize different measurement conditions. The isometric lines are drawn through the interval equal to standard deviation. Each distribution

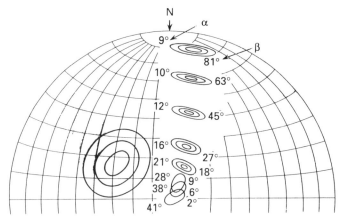

Figure 5.2 Bivariate error distributions when measuring fracture orientation on the Schmidt net. Mathematical expectation of measure valued is denoted by figures. Isolines correspond to errors σ, 2σ, 3σ. For comparison see natural distribution of bedding fracture orientation in travertine, Mt Mashuk (left). As is seen, its dispersion far exceeds measuring error

is characterized by one standard deviation related to the error in measuring the dip angle β and the other standard deviation related to the error in measuring dip azimuth α. Standard errors calculated for concentration ellipse axes are tabulated in Table 5.2.

Table 5.2 Standard errors in measuring fracture position with surveyor's compass at different angles

Dip angle (degree)	2	6	9	18	27	35	45	55	63	81
Standard S errors (degree)	0.8	1.2	1.3	2.1	2.7	–	3.0	–	3.2	3.1
S	1.6	1.6	1.5	1.5	1.3	1.0	0.9	0.8	0.8	1.1
Number of measurements	100	100	160	100	140	40	80	20	80	50

Parallel measurements of fracture arrangement to assess methodological errors have been taken on outcrops of Paleogene limestones in the middle course of the Vaksh river in Tajikistan. The arrangement of one and the same fractures has been determined by two geologists (Table 5.3).

Table 5.3 Assessment of the position of fracture set centre by two operators

Operators	Number of samples	Average dip azimuth and standard deviation (degree)	Average dip angle and standard deviation (degree)
1	35	75.5 ± 5.3	62.0 ± 4.0
2	35	73.0 ± 5.3	64.0 ± 3.5
1	50	63.7 ± 2.5	62.6 ± 1.2
2	56	68.3 ± 1.7	61.6 ± 1.1

Measuring accuracy is largely determined by the skill of the operators. Table 5.4 lists the results of measurements taken by operators with similar experience, whereas Table 5.6 presents data obtained by two operators whose experience and skill differ greatly. Measurements have been taken on one areally limited outcrop. The total number of measurements in each of the series was 200. All in all, there were at least 2000 fractures that could be measured on the outcrop. Thus, the series of measurements overlapped only in part. As is evident from the comparison, the results of measurements are close to each other, with a minor statistical difference. Hence, there is no need for special training to survey fractures.

Fracture width or aperture is measured to determine porosity, permeability and deformability. For taking measurements, use is made of a feeler gauge or metal rule with a magnifying glass. As an alternative, fracture photography followed by strong magnification can be employed.

The measuring error was investigated with the aid of a rule and a magnifying glass under laboratory conditions. A clearance between the vernier caliper legs was taken as a fracture model. It was measured with a metal rule through a magnifying glass and then with the help of vernier reading. A total of 111 measurements were taken. The standard error was 0.06 mm.

Table 5.4 Assessment of the position of fracture set centres based on measurements with error of analogy

Operators	Number of samples	Average dip azimuth and standard deviation (degree)	Average dip angle and standard deviation (degree)
1	20	21 ± 0.6	35 ± 0.6
2	20	21 ± 0.7	37 ± 0.7
1	83	119 ± 0.8	79 ± 0.8
2	80	116 ± 0.7	84 ± 0.7
1	22	153 ± 2.0	51 ± 2.0
2	17	150 ± 3.0	62 ± 2.0
1	16	267 ± 2.5	66 ± 1.0
2	20	266 ± 2.2	66 ± 2.2
1	19	357 ± 1.5	76 ± 2.0
2	10	3 ± 3.0	74 ± 3.0
1	33	209 ± 1.5	51 ± 2.0
2	31	209 ± 1.5	57± 1.5

Table 5.5 Assessment of accuracy of fracture width measurements by photography

Sample no.	Average fracture width (mm) (based on measurements)	
	radioactive-tracer method	special photographs
1	0.34	0.24 ± 0.024
2	0.30	0.27 ± 0.034
3	0.29	0.30 ± 0.040

Table 5.6 Results of measuring fracture width by different operators

Adit no.	Fracture no.	Fracture width (mm)	
		1	2
777	17	20.0–45.0	12.0–36.0
	19	0.5–1.5	0.4–4.0
	25	0.5–1.5	0.4–1.1
	52	2.0–10.0	3.0–9.0
763	4	0.0–8.0	0.2–1.7
	17	0.2–3.0	0.2–1.7
	20	0.0–3.0	0.2–1.2
	37	0.0–1.5	0.2–2.0
	3	0.5–10.0	0.1–1.2
	26	0.0–3.0	0.4–4.1
	80	10.0–15.0	27.0–174.0
	41	0.1–1.5	0.1–0.8
	53	0.1–0.2	0.3–1.2
	45	0.1–2.0	0.1–3.7

The methodological error related to schematizing of individual fracture width was studied on granite core samples. Two methods were employed for measuring fracture width under laboratory conditions (Table 5.5). The total number of fracture width measurements was 725. The two methods yielded results fairly close to each other statistically, the absolute error not exceeding 0.1 mm.

The methodological error in measuring fracture width was also studied in the Koyma Hydro adits. In this case use was made of a feeler gauge. Table 5.6 lists maximum and minimum values of fracture width. The results of these measurements show good agreement.

Fracture spacing

Fracture spacing is measured along the perpendicular to the fracture surface using a folding rule. Sections ranging in length from 2 to 5 cm to 2 to 5 m are measured. The error in measuring such values is insignificant. To prove this data were obtained by parallel determination of fracture spacing in Palaeogene siltstones in Tajikstan on the outcrop and from a pair of stereoscopic photographs taken from a distance of 5 m. The results obtained by the two methods are statistically similar. For the outcrop the average spacing and standard deviation were 26.0 and 15.4 cm respectively. From the photographic measurements these figures were, respectively, 26.0 and 15.5 cm. The standard error in measuring an individual fracture in this case was 1.44 cm with a total of 136 measurements taken. The relevant variation coefficient is 5.5%.

Fracture length

The results of fracture length measurements may apparently differ only because of the selection of a fracture trace termination on an outcrop or due to the rule reading error. These errors are generally insignificant.

To prove this, information on fracture length measurements taken on the same outcrop where fracture spacing was determined may be analysed. A total of 101 fractures were measured lengthways on the same outcrop and on the photograph. There is some uncertainty that the same fractures were measured in these two cases. However, there is no doubt that they belong to the same set. For the outcrop the average length of fractures was 322.0 cm with 310.4 cm standard deviation. For the photographic measurements these figures were, respectively, 347.0 and 293.6 cm.

Also assessed was the error in measuring fracture length due to surface roughness in plan. Ten fractures on marble samples were measured lengthways with a rule. The same fractures were measured on enlarged photographs. After comparing the results of these measurements, the error is 5.5% due to increased length of fractures stemming from their roughness in plan.

Comment

Thus it has been shown that the errors in measuring fracture parameters are insignificant as compared with methodological and interpolation errors. The error in measuring fracture length and spacing is generally about 5%. The accuracy of fracture width measurements is lower. In the inner portions of a rock mass where the average fracture width is about 0.5 mm, the relative error is between 30 and 50%. In zones of increased fracturing where fractures are much wider, the relative error in measuring fracture width is 10%.

The errors in measuring dip angle and azimuth (see Figure 5.1), if taken individually, fail to characterize the accuracy of determining the spatial position of a fracture. It is only in combination that one can assess the error in measuring fracture orientation. The mathematically rigorous study of bivariate error distribution suggests that the accuracy of determining the spatial position of a fracture does not depend upon dip angle and is characterized by standard deviation of individual measurements within 2–3° (see Figure 5.2).

The measuring errors can be judged only through comparing those with standard deviations of natural distributions of a measured parameter (Rebrik and Chernyshev, 1968) since it is not absolute errors that matter, but their contribution to the general dispersion of results of measuring a natural random value. In most cases the aforesaid errors in measuring fracture parameters are much lower than natural distribution standards (see Chapter 3). On the whole the accuracy of measuring fracture parameters in engineering–geological surveys is satisfactory.

5.4   Photogeodetic and borehole studies

Geodetic methods are employed for studying fracture systems on hard-to-reach outcrops and open-pit edges (Famitsyn and Fedorenko, 1970; Vieten, 1970; Ross-Brown and Atkinson, 1972; Ross-Brown, Wickens and Markland, 1973; Moore, 1974; Preuss, 1974). Use is made of stereophotogrammetry with phototheodolites and cameras with subsequent processing of photographs on special instruments such as a stereometrograph, a stereocomparator, or a stereoautograph. As a result adequate information on fracture systems can be obtained. For some large fractures dip angle and azimuth can be determined (Vieten, 1970). Fracture length and spacing and the location of zones of increased fracturing can also be assessed (Franklin, 1986, 1988).

The photogeodetic study is a complex process calling for highly-skilled personnel equipped with appropriate instruments. It is advisable to employ photogeodetic techniques in cases where hard-to-reach outcrops are involved or when there is lack of time for field work and the emphasis is, therefore, placed on desk studies.

For the description of fracturing based on drilling records, the borehole walls and the core samples are available. A great deal of attention has been given to methods of core drilling to give a high recovery of undisturbed core. This can best be achieved using diamond drilling with double or preferably triple core barrel with a split inner barrel and a non-rotating rigid plastic liner. In difficult ground, and using a foam drilling fluid, samples can be recovered with original joint fillings intact.

The problem of core orientation remains if the dip azimuth of the discontinuities is to be determined, and a number of elaborate systems have been devised.

If marker horizons can be identified in a core run, and at least three suitably distributed boreholes are available, then the well known three-point graphical solution of the problem can be applied.

For a single borehole, some form of orientation device can be used as the holes are drilled (Hoek and Brown, 1980). The Atlas Copco–Craelius core orientation system uses a tool which is clamped in the core barrel as it is lowered into the hole at the start of drilling. A number of pins parallel to the drill axis project ahead of the drill bit and take up the profile of the core stub left by the previous drilling run. The orientation of the device is determined relative to the drill rod position at the collar of the hole. The first piece of the recovered core is matched to the profile of the pins, and the remainder of the core is carefully assembled to obtain the orientation of other discontinuities relative to the first piece of core.

A technically more complicated method involves drilling a smaller diameter hole at the end of the hole left by the previous core run. An oriented rod can be grouted into the hole to provide reinforcement for the core as well as orientation (Rocha, 1967). The method is both expensive and time consuming but it can be used to produce high-quality oriented core in very poor rock, important in the evaluation of extremely critical areas in the rock mass.

By contrast, theoretically, all the parameters can be obtained from the borehole walls. However, they are difficult to reach and call for sophisticated television scanning devices and camera systems (Wittke, 1984). The results are seldom satisfactory especially if the water in the hole is dirty.

An inexpensive tool has recently been developed which gives an impression of the inside of a diamond drilled hole (Hinds, 1974; Barr and Hocking, 1976). A special thermoplastic film is pressed against the

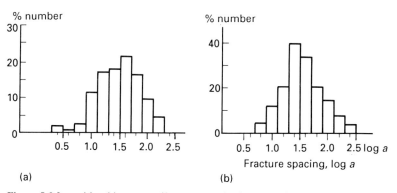

(a)   (b)

Figure 5.3 Logarithm histograms (fracture spacing in traps of the Tolstomysovsky intrusion for the horizontal fracture set) (a) for outcrops; (b) for core extracted from hole drilled close to the outcrop

borehole wall by an inflatable rubber packer. Information on the orientation of fractures in the rock mass, independent of disturbance of the core,
is thus provided if the device is linked to some form of borehole orientation
system or borehole survey instrument.

Experience has shown that with high core recovery in hard rocks, say
quartz porphyry or traps, a reliable assessment can be made of the spatial
position of a fracture or the fracture spacing in a set (Figure 5.3). However,
it is sensible to attempt to relate parameters derived from cores or
boreholes to the rock mass by examining natural exposures or large
diameter exploratory openings.

Chapter 6
Desk processing of field measurements of fracture parameters

6.1 Spherical coordinate grid projections used in plotting structural diagrams

For graphical processing of fracture orientation measurements, use is made of spherical coordinates, and more specifically their projection on a plane.

The ideal spherical coordinate system is assumed to be on the sphere surface. A trace of the plane passing through the sphere centre is referred to as a great circle. Points on the sphere surface equidistant from the great circle are referred to as poles. Two poles correspond to each great circle or equator.

The spherical coordinate systems are used to determine the position of a point on the sphere. To plot the coordinate system on the sphere, two diametrically opposite points are fixed as poles and a great circle (equator) is drawn. Great circles passing through the poles are termed 'meridians'. One of those is taken as the origin. As for the others, their position is determined by the angle between the base meridian and the plane of the meridian involved. The position of a point on the sphere is determined by two angles: longitude α measured between the zero meridian and local meridian and latitude β measured with the meridian arc from the equator to a given point.

Plotted on the sphere are also parallels, circular sections parallel to the equator. All the points of the parallels have the same latitude. A set of meridians and parallels drawn through a particular interval forms a spherical coordinate grid. The projection of the spherical coordinate grid on the plane is referred to as a 'cartographic grid'.

The term 'projection' implies any representation of the sphere on the plane, with point M on the plane corresponding to each point M_0 of the sphere. The movement of point M on the plane along the line which is a meridian projection corresponds to that of point M_0 of the sphere along the meridian. In summary it may be said that a geometrically determinate point, line or figure on the sphere projection corresponds to the point, line or figure on the sphere.

When studying structural patterns of fracture sets, use is made of azimuthal projections of spherical coordinates as polar and equatorial grids (Figure 6.1). Both polar and equatorial grids feature projections of a hemisphere. **Projections of the upper hemisphere are employed in Soviet practice**, whereas those of the lower hemisphere are typical of other authors (Phillips, 1971; Hoek and Brown, 1980). The term 'polar grid'

Polar   Equatorial

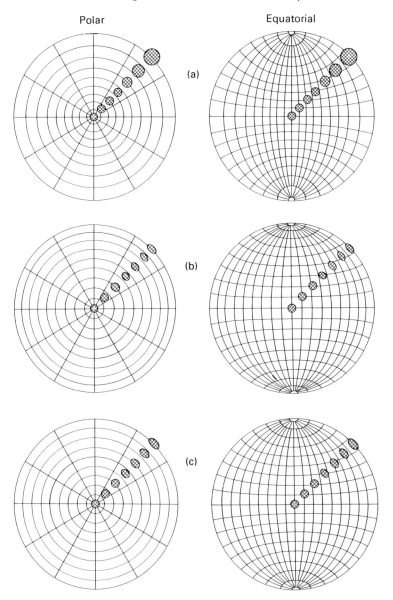

Figure 6.1 Equatorial and polar cartographic nets: (a) Wolf stereographic net; (b) Schmidt aquiareal net; (c) equidistant net. Cross-hatching shows projection of circle of unit radius onto plane (circle projected from different parts of the sphere).

implies a grid in whose centre there is a pole and the equator is represented by the grid circumference. Meridians are shown by radial lines and parallels by circles. In the equatorial grid poles are located on its circumference and the equator is shown by a diametral straight line, the perpendicular line connecting the poles.

Mathematically, the idea of constructing azimuthal projections is as follows. For the polar azimuthal projection the longitude α on the plane is always equal to the longitude $α_o$ on the sphere. In other words, the angles between the meridians on the polar projection are shown without distortion. The latitude β on the cartographic grid is shown by the distance $P = f(β)$. This function may be different in form, each form featuring a particular type of projection. Three types of azimuthal projections are employed in structural analysis in geology, namely (i) equiangular or stereographic introduced into crystallography by Wolf; (ii) equiareal introduced into structural geology by Schmidt; and (iii) equidistant.

To construct an equiangular or stereographic projection, the function takes the form

$$p = k \times \tan \frac{v}{2}$$

where

k is a distance from the sphere centre to the plane on which the sphere is projected;
v is polar distance;

$$v = 90° - β$$

In Wolf's equiangular grid the coefficient is taken to be equal to the sphere radius. With the function given like this, the angles between the curves at each point of the projection are equal to the sphere angles. Therefore, circles on the sphere are shown as those on the projection. Distortions affect figure areas. The area of the circle in the projection centre (Figure 6.1) is equal to that on the sphere, whereas in the outer contour of the grid the area of the circle on the projection is four times the area of the circle projected from the sphere. Wolf's grid with equiangular projection can be used when measuring angles between points or passing sections through a coordinate sphere with a plane or cone, for example the friction cone in solving slope stability problems. The grid is not suited to statistical processing of large-scale measurements of fracture orientation since the density of points on the projection is largely determined by their coordinates.

In the event of equiareal projection, the main equation takes the form

$$p = 2R \sin (v/2)$$

where R is sphere radius.

In this case the areas of figures on the sphere and projection are equal. However, arc lengths and angles, when transferred from the sphere onto the projection, are distorted. Hence circles, except for the one in the grid centre, are represented on the equiareal projection as ellipses.

Equiareal azimuthal projection grids proposed by Schmidt are used in plotting structural fracture diagrams. Their main advantage is a constant density of the distribution of points both on the sphere and on the projection. This feature of the grid enables point density isolines to be drawn using a simple graphical technique.

Of the projections discussed above, one projection distorts angles and does not affect area. By contrast, the other distorts area and does not affect angles and the form of figures. Equidistant projection helps reduce all these distortions. However, in this case both angles and areas are distorted simultaneously. Type of grid is of no importance in computer calculations of point density using templates developed by Pronin or Braich (Vistelius, 1958). In this case one can use the equidistant projection, which is recommended for computer processing (Section 6.3).

6.2 Representation of straight lines and planes on a circular diagram; plotting the diagram

In the practical analysis of fracture and fault patterns, some objects are represented as a plane. These include fractures, faults, surfaces of strata and outcrops. There are other objects (e.g. fold axes) shown as a line.

First, consider the representation of a straight line and plane on the coordinate sphere (Figures 6.2 and 6.3). Consider the straight line (left) inclined at the angle β with the horizon and rising in the direction of the azimuth α. It is not the line proper but its projection on the horizontal plane that has the azimuth α. Choose such a position of the coordinate sphere so that the line passes through the centre of the sphere. The line will intersect the sphere at the point A_0. The equatorial plane of the sphere is horizontal. In this plane plot the line rise direction azimuth from the origin. Pass the meridian plane through the point obtained and plot the angle β in this plane, so the point A_0 is obtained. To transfer the point A_0 onto the sphere projection, it is necessary to plot the radial distance p in the direction α from the centre of the diagram on the equatorial plane. Referring to Figures 6.2 and 6.3 $p = 0A$. The value p is determined by the equation $p = f(\beta)$. However, practically there is no need to refer to the equation since the radius is broken down into intervals on the scale β.

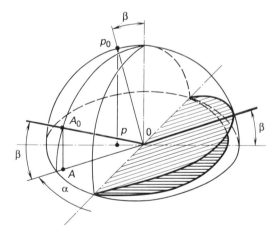

Figure 6.2 Representation of point and plane on coordinate sphere and their projections on to the equatorial plane

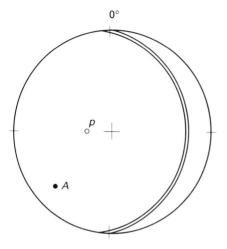

Figure 6.3 Equatorial plane (top view)

Therefore, to find the point A, a segment has to be counted off equal to β from the great circle to the centre of the diagram.

A plane is shown on the diagram in two ways, namely as a circle (circular arc) or a point. Meridians and parallels are shown on a cartographic grid as circles (circular arcs). Intersection of a plane with a coordinated sphere produces a circular arc. To plot the projection of the plane as a line, it is necessary to draw a meridian on the cartographic grid which is separated from the great circle by the angular value β.

To represent a plane as a point, one has to construct a perpendicular to the plane from the centre of the coordinate sphere. The perpendicular intersects the sphere at the point p_o; the projection p of this point on the equatorial plane is an adequate representation of the plane on the cartographic grid. Showing a plane as a point allows representation of several hundreds of differently oriented planes on one and the same diagram. This method is, therefore, widely used when plotting circular diagrams of fracturing.

6.3   Computer processing of measurements

When large numbers of structural geology data are involved, the desk study now tends to the use of the computer (Spencer and Clabaugh, 1967; Ivanova and Chernyshev, 1974). The integrated program incorporates the following procedures:

- plotting of a circular diagram of fracturing, including determination of point distribution density and isolation of fracture systems;
- calculation of fracture system statistics;
- computation of general fracture characteristics, namely porosity and block size.

Processing of the results of large-scale measurements of fracture parameters is aimed at defining general trends of fractures, determining such parameters as average spacing, width and length, and providing an insight into the main generalized parameters of fracture sets, namely porosity and block size.

Data on fracture systems in natural or artificial (pits, mines, adits) exposures are to be processed. For each particular target area a total of 200 fractures is taken for measuring dip angle and azimuth. Then 3–5 main fracture systems are identified and 30–50 such parameters as fracture spacing, opening and length are measured in each of those. All these data are provided in tabulated form.

The first processing operation involves plotting of a circular fracture diagram, which is generally plotted in accordance with Schmidt's method. This method, developed in the 1930s, had a major role to play in structural studies, yet it has a number of shortcomings. As recommended by Vistelius (1958), the equidistant sphere projection (Kavraisky grid) and Braich's templates are used as a coordinate system to determine density of points. Data loaded into the computer comprises 200–1000 measurements of fracture dip angle and azimuth. Figure 6.4 illustrates circular diagrams plotted manually and with the aid of the computer on the basis of 200 measurements.

The average fracture spacing, dispersion and standard deviation of fracture spacing are computed as part of the desk study. These computations are performed for each individual fracture set. Such parameters as average width and length, dispersion and standard deviation are computed in the same manner for each particular fracture set.

Porosity

The average values of fracture spacing and width thus obtained are used in computing porosity (n), namely the percentage ratio of fracture volume to the volume of the rock mass involved:

$$n = \sum_{i=1}^{n} \frac{b_i}{a_i + b_i}$$

where
  a_i = average fracture spacing;
  b_i = average fracture width.

This expression does not include second- and third-order infinitesimals allowing for intersection of fractures. With large b and small a when $n > 0.05$ a systematic positive error is allowed. The calculated value proves to be higher than the actual one. This error makes up a few percent of the result and is directly proportional to fracture width. With allowance made for fracture intersections the rock mass porosity is the sum total of porosities in sets minus the porosity of intersections in each of the two sets plus the porosity resulting from the joint intersection of each of three sets. The probability that more than the three fractures will intersect at the same point is assumed to be zero. For these conditions the formula of rock mass porosity with five (or less) fracture sets is as follows:

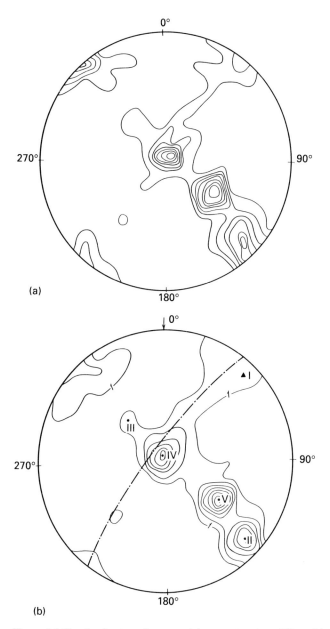

Figure 6.4 Circular fracture diagrams: (a) computer-plotted Kavraisky grid; (b) Schmidt net plotted by Schmidt's method

$$n = \sum_{j=1}^{5} n_j - n_1 n_2 - n_3 \sum_{1}^{2} n_j - n_4 \sum_{j=1}^{3} n_j - n_5 \sum_{j=1}^{4} n_j + 2 \left[n_1 n_2 n_3 + n_4 (n_1 n_2 \right.$$

$$\left. + n_2 n_3 + n_1 n_3) + n_5 (n_1 n_2 + n_1 n_3 + n_2 n_3 n_1 n_4 + n_2 n_4 + n_3 n_4) \right]$$

where $n_1, n_2, \ldots, n_j$ is porosity due to individual fracture sets Nos 1, 2, 3, 4, 5.

The contribution of each particular set to the total porosity of the rock mass is

$$n_j = b_j / (b_j + a_j)$$

where

    b_j is average fracture width;
    a_j is average fracture spacing.

Block size

Block size, much like porosity, is also regarded as the main classification factor of a fractured rock mass. Tables 6.1 and 6.2 offer a classification of rock masses on a basis of porosity and block size with an indication of modulus of deformation and permeability coefficient. Block size can be determined through various techniques, one of those being proposed by Kolichko (1966). This method enables the distribution of rock blocks to be obtained as a histogram coinciding with the one based on block measurements in a spoil bank following blasting.

Table 6.1 Calculated values of permeability coefficient (m/day) for fractured rock mass at different porosity and block size

Porosity (%)	Block size across (cm)		
	300–100	100–30	30–10
0.1–0.3	0.1	0.01	0.005
0.3–1.0	5.0	1.0	0.05
1.0–3.0	100.0	10.0	5.0
3.0–10.0	–	500.0	50.0

Table 6.2 Ratio between rock sample deformation modulus E_R and rock mass deformation modulus E_m at different porosities

Porosity (%)	E_R/E_m ratio
0.1–0.3	2
0.3–1.0	5
1.0–3.0	20
3.0–10.0	50–100

Kolichko's principle was used in computer calculations of block size. His method, as well as the one developed by Schmidt to plot circular diagrams, is based on a graphical procedure. A graph field is divided into polygons with sides equal to actual fracture spacings on an appropriate scale. The area of the polygons is calculated. Each value thus obtained is multiplied by the average thickness of a layer to produce a totality of rock block volumes essential to a histogram. The sought distribution of block size in a rock mass is a function of fracture spacing distributions. In the event of graphical determination of the block size distribution functions some errors are allowed which are partly inevitable due to the graphical realization of the method:

1. The distribution of fracture spacings in one set referred to as that of bedding fractures is assumed to be constant based on the average spacing. This assumption leads to the distortion of the sought distribution, reducing in particular its dispersion. The number of very large and small blocks diminishes and that of medium-sized blocks increases. The error is minor when studying stratified formations with a small dispersion of stratum thickness. In cases where igneous and sedimentary rocks with a large thickness dispersion are involved the error may be great.
2. In the graphical analysis there are some blocks with a side ratio of 10 : 1 and up. Such blocks may exist in a rock mass, where in a spoil bank they will be broken down into smaller ones. The graphical method fails to take account of block breaking and hence it should be regarded as the one for assessing block size in a rock mass. To determine block size in a spoil bank, an additional operation of block breaking is required.

The above errors have been eliminated in the computer program for assessing block size. The program is aimed at simulating rock block samples with volumes $V_1, V_2, \ldots, V_n$. The volume of each ith block $V_i = f(a_{i1}, a_{i2}, a_{i3}; P_{I-II}, P_{II-III})$.

The tetrahedron volume is the product of the area of the section perpendicular to one of the edges by height. The area of the section perpendicular to the edge formed through the intersection of sets 1 and 2 is in fact the area of a parallelogram with the apex angle equal to the angle between fracture planes 1 and 2. Height is fracture spacing in system 3 with a correction, provided the angle of intersection of said edge with set 3 is not equal to 90°. As is evidenced by studies of actual fracture systems (see Chapter 2), there is always a pair of orthogonal sets. These are assigned numbers 1 and 2. In this case

$$v_i = a_{i1} a_{i2} a_{i3} / \sin P_{I-II}$$

where

a_{i1}, a_{i2}, a_{i3} are spacing between ith and $(i+1)$th fractures in systems 1, 2 and 3 respectively in cm;
P_{I-II} is the angle between the edge of intersection of systems 1, 2 and system 3 in degrees.

The total number of blocks following the multiplication of spacings a_1, a_2, a_3 is equal to the product $n_1, n_2 n_3$ of the number of spacings measured in sets 1, 2 and 3. In this case each of the series is a sequence of random values and the first error of the graphical procedure is corrected. The number of terms of the computer-simulated sample increases generally by 30–50 times as compared to the graphical sample (30–50 is the number of fractures measured in system 1, referred to as the system of bedding fractures). Thus a matrix of the order of $n \times 4$ is formed:

$$\begin{vmatrix} a_{11}\ a_{12}\ a_{13} \cdots & V_1 \\ a_{21}\ a_{22}\ a_{23} & V_2 \\ \vdots & \\ a_{n1}\ a_{n2}\ a_{n3} & V_n \end{vmatrix}$$

To correct the second error, find a_{max} and a_{min} in each row of the matrix and calculate the relation a_{max}/a_{min}. To determine the limiting value of this relation, the condition

$$a_{max}/a_{min} \leqslant F$$

is to be met, where F is maximum proportion of edges at which a block is not broken down when broken off the solid. One may take $F = 5$ for isotropic rocks and $F = 10$ for anisotropic rocks (schists, etc.).

Form a row vector (V_i) as follows. Should the last condition be met for the respective matrix row, enter the respective value V_i in the row. If the condition is not complied with, calculate

$$V_i^i = V_i/F'$$

where $F' = [a_{max}/F\ a_{min}]$. In this case the value V_1^i is entered in the row vector F' times.

Tables 6.3 and 6.4 list results of 'Block' program computations. Table 6.3 presents data on limestones exposed in the Nurek Hydro area. Block size measurements following the blast were taken by geologists from the Hydroproject Research Institute. Fracture measurements were taken by the authors. Table 6.4 sums up the application of the program to assess

Table 6.3 Statistical distribution of rock block volumes obtained by different methods

	Block size assessment methods, per cent		
Block volume m³	measuring blocks in spoil bank after blast	graphic processing of field measurements of fracture spacing (after Kolichko)	computer processing of field measurements of fracture spacing
0.0 –0.02	66.2	63.0	71.8
0.02–0.05	16.5	14.0	15.4
0.05–0.07	11.8	15.0	5.0
0.07–0.09	3.9	4.0	2.6
0.09–0.11	1.6	4.0	2.6
0.11–0.19	0	0	0
0.19–0.21	0	0	1.3
0.21–0.35	0	0	0
0.35–0.37	0	0	1.3

Table 6.4 Rock block size

	Intervals		Distribution of blocks in intervals (%) at survey points			
No.	block size across (cm)	block weight (kg)	453	466	457	452
1	0–5	0.2	0	17.1	0	0
2	5–10	1.5	36.7	45.7	5.7	33.3
3	10–15	5.9	50.5	34.3	14.3	36.4
4	15–20	15.3	13.3	2.9	25.7	27.3
5	20–23.3	27.8	0	0	14.3	3.0
6	23.3–30.0	53.5	0	0	28.6	0
7	30–35	94.3	0	0	5.7	0
8	35–40	114.3	0	0	2.9	0
9	40–45	209.4	0	0	2.9	0
10	45–50	291.8	0	0	0	0

rock block size in the area of the Erdent Mining and Concentration Complex in Mongolia covered by an engineering–geological survey.

Experience has shown that computer processing of fracturing data saves a lot of desk study time without detriment to quality.

Chapter 7
Indirect methods for studying rock fractures

A system of fractures in a rock mass has some definite parameters, such as orientation, density, width and length. Direct measurements of these parameters enable a structural model to be constructed of a fracture system for particular key areas in the rock mass and study variation of the parameters when passing from one area to another. The model, based on direct measurements of geometrical parameters, helps to solve a wide range of practical and research problems. However, direct measurements often prove non-feasible because of lack of outcrops or exploration openings. So along with direct measurements of fracture geometry, provision is also made for indirect methods based on the relationship between fracture parameters and rock mass properties.

Indirect assessment of fracturing involves loss of data. Fractures actually affect all rock mass properties. No process occurs in a rock mass without reference to fracturing. So there is a great variety of indirect methods. However, not a single process and, apparently, not a single property of a rock mass are determined by some particular parameter of a fracture system. These properties and processes are governed by a combination of fracture and rock parameters. Indirect methods generally allow rough assessment of fracturing in different areas of a rock mass. They are widely used in rock mass zoning and in studying water permeability and other properties.

The most common indirect methods are based on the determination of water or air absorption by a fractured rock mass or on the measurements of elastic wave velocities. Their main advantage is that they provide an insight into the deep-seated portions of a rock mass that are inaccessible to direct investigations.

7.1 Geophysical methods

To detect and trace zones of increased fracturing, determine the trend of fracture sets and rock mass void ratio, use is made of seismic exploration, electrical prospecting, emanation survey and magnetometry in combination with direct fracture measurements. Integration of geophysical methods seem to be most promising.

Seismic methods are employed to determine fracturing intensity and variation. Assessment of small fractures is basically similar to that of fault zones. To determine trends of major fracture sets, elastic vibrations are

156

used, wavelength being of the order of fracture spacing. Ultrasonic frequencies (10–20 kHz) are generally employed.

Results of studies made by A. I. Savich in the Inguri Hydro area may serve as an example of effective assessment of fracture sets using seismo-acoustic techniques. In the monoclinal sequence of Lower Cretaceous limestones on the Inguri River (Western Georgia), rocks are characterized by intense tectonic fracturing. Geologically, a total of six fracture sets was singled out, which are associated genetically with local folds and faults. There is a large strike slip fault within the dam foundation area. Fractures are fairly wide, most of them being 1 cm wide. The width of 25 fractures exceeds 10 cm. The fractures are filled in with clay breccia, clay or are healed with calcite. Specific water absorption is heterogeneous. Side by side with low water absorption zones, there are zones featuring high water absorption accounted for by intense fracturing. Ultrasonic observations with an observation base of 25–30 cm were carried out to define the trend fracture of sets. P-wave velocities V_p were obtained for 10–16 different trends. The wave velocities vary greatly with trend (Figure 7.1) reaching 6 km/s for trends along fractures and crush zones, whereas at right-angles they are 2.0–2.5 km/s. Interpretation of the vector diagram is confirmed by comparison with the fracture rose diagram.

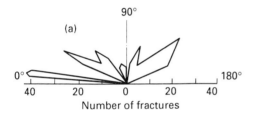

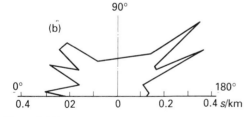

Figure 7.1 Fracture rose diagram (a) and vector diagram of propagation of ultrasonic vibrations (b) for limestones, Inguri Hydro area

The rock mass porosity was also determined by A. I. Savich who measured propagation rates of low-frequency oscillations. Along with seismic profiling, a detailed engineering–geological survey was carried out. The porosity turned out to be 4.5–0.45% based on seismic exploration data, whereas direct measurements gave another figure (3.0–0.3%). Seismoacoustic measurements covered the inner portions of the rock mass, whereas geological investigations involved exposed rocks on the adit walls with expanded fractures. So it should be noted that the rock mass porosity as determined by seismic methods is in error of at least two times on the high side. Yet such errors are allowable in the rough assessment of porosity.

Seismic profiling was carried out in the Moscow area (Lyakhovitsky, Morgun and Chertkov, 1981) to identify zones of increased fracturing in Carboniferous limestones. High-frequency filtering (70–100 Hz) refraction profiling was employed. The boundary velocity at the top of carbonate rocks was 2.6–4.1 km/s. The zone of increased fracturing was identified by the following factors: (i) lower boundary velocity; (ii) different wave pattern (elimination of converted waves, type PSP); (iii) lower amplitude and apparent frequency of vibration; (iv) complicated form of refracted wave recording due to interference. The main evidence of zones of increased fracturing was a boundary velocity which characterizes the upper 20 metres of the limestone unit. As a result, it has become possible to identify areas with different fracture patterns.

Thus, seismic exploration methods allow a rough assessment of fracture parameters. In other words, they make it possible to assess rock mass heterogeneity and anisotropy due to fracturing.

A complex of geophysical methods for tracing faults and zones of increased fracturing under Quaternary deposits was used in seismic zoning of the industrial area of Erdenet in Mongolia (Anon., 1980). The complex included emanation survey, seismic exploration, electrical and magnetic prospecting. The engineering–geological operations were performed with such methods as airborne interpretation, serial and ground observations, and core drilling.

The area is composed of granodiorite, granite, porphyry and other igneous rocks which are largely metamorphosed and faulted. Many faults with displacements of hundreds and thousands of metres were formed in Cenozoic times and are active in the recent period. The earthquakes recorded in the area had a maximum magnitude of 10. Clearly defined seismotectonic dislocations can be observed, the youngest of which are associated with the 1967 Mogod earthquake.

Use was made of a CMII–24 seismic station with signals recorded on a ferromagnetic tape and subsequently represented on paper. The operations were performed with the refraction method, continuous seismic profiling along survey lines or seismic sounding at individual points. Twenty-four geophones (C–120 or C–130) mounted on a 46-m-long cable were used as sensors. The geophone spacing was 2 m. At each station recording was performed from four source points with reversed and catching-up TD curves obtained at 0–46 and 46–92 m from the source point, namely the maximum length of the time–distance curves reached 92 m. The elastic vibrations were generated with a 8–10 kg hammer hitting a wooden plate buried into the ground with a blow. The operations performed in the territory of the town of Erdenet involved the use of an impact unit, 5 m geophone spacing and a maximum time–distance curve length of 230 m. The operations were completed prior to the construction stage.

Most of the seismic exploration data were obtained using the ZZ observation system: vertically directed impacts and vertically arranged geophones. For some 20–30% of lines and soundings, the YY observations were carried out, namely horizontally directed impacts and horizontally arranged geophones. The combined observations (systems ZZ and YY) enable reliable identification of both S- and P-waves.

Processing of seismic data involved plotting of time–distance curves and determination of boundary velocities and boundaries. To determine average velocities in individual strata or in the drift unit as a whole, special borehole investigations were made. The seismic lines on the construction sites coincided with some profiles along which holes were drilled or openings cut. The seismic exploration program involved a system of profiles run in the main project areas. Individual seismic soundings, logging and measurements of average velocities were made for most of the holes. Seismic exploration was performed on seismic dislocations in the Erdenet, Mogoingol and Mogod areas. Seismic profiles were run at an average rate of 0.5 km per 1 km^2 of the prospect area. On the construction sites each 1 km^2 of the area had 2–4 km of profiles.

Electrical prospecting

A standard compensator, type 3CK–1, for DC electrical prospecting was used. The modifications included vertical electric sounding (VES) and symmetric and non-symmetric electric profiling.

Vertical electric sounding was employed for dissecting a rock unit into geoelectric layers and mapping the top of the rock. The maximum dimensions of the VES array reached 2000 m, thus providing an insight into depths of up to 100 m. The VES points were generally arranged along the drilling survey lines with a spacing of 25–30 m along the profile. Since boreholes were spaced 100–200 m along the profile and their depth did not exceed 10–20 m, failing to reach rockhead, the VES data provided valuable information on the cross-section. Parametric electric sounding was made on all the holes drilled to a depth of less than 100 m.

Symmetric and non-symmetric electric profiling was aimed at identifying fracture and crush zones in rocks. Two to three spreads were used. The profiling was detailed and the point spacing, therefore, varied from 1 to 10 m. The symmetric and combined profiling graphs (p_k) feature maxima generally associated with near-surface fractured and crush zones with air serving as the interstitial filler (Figure 7.2).

Emanation survey

Emanation anomalies are accounted for by the processes of gaseous emanation. This is a complex phenomenon based on alpha decay of radium, thorium and actinium isotopes. The migration of rock emanations is caused by diffusion and convection. Until recently it has been presumed that the emanation anomalies occur in deposits overlying radioactive ore bodies or large fault zones, the overlying thickness not exceeding 8–10 m. Only radon (decay period 3.82 days) emanations were considered to be of importance. Thoron and actinon decay periods are, respectively, 54.5 and 3.9 days. Thorium and actinium decay at an extremely rapid rate and their migration potentialities are low. Based on these considerations, it is not

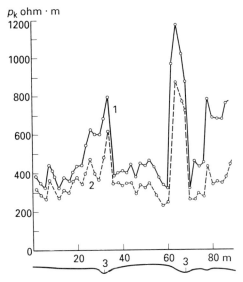

Figure 7.2 Electric profile across seismic dislocations (Anon., 1980): (1) and (2), symmetric and non-symmetric profiling; (3) axes of small gullies at the ground surface.

possible to account for the origin of intense emanation anomalies (mostly thoron) in thick (up to a few hundred metres) cover beds with no large faults, radioactive ores, convectional jets, etc.

Based on practical experience in the Donetsk coalfields and laboratory tests, such anomalies may be attributed to the fact that modern medium- and small-amplitude movements result in a dramatic increase of radio-active emanation contents (Gorbushina and Ryaboshtan, 1974), emana-tions being inherent in the cover proper. Presently there is a version on space–time relationship between emanation anomalies and zones of recent tectonic movements.

The recent tectonic movements give rise to new and revive old faults in a rock mass. It is fairly difficult to identify such faults under the overlying strata, yet emanation survey integrated with other geophysical methods helps to solve the problems.

The emanation survey involved air sampling from depths of 50–70 cm followed by analysis in an ionization chamber. The point spacing along the profile was 5–10 m, the profiles being arranged across anticipated faults. Emanometers, type M6 and radon, were used. The results were expressed in terms of the number of counts per minute. The emanation background in the Erdenet area is 100–120 counts/min.

The anomalies attributed to the effect of geodynamically active zones exceed background contents by 3–5 times. They are essentially of a thoron character, a proof of which is a 2–3 times decrease in their intensity during the first 2–3 minutes.

The intensity of anomalies is apparently related to the state of a fault zone at each instant of time. Studies made in Erdenet along the same lines show that the emanation intensity in anomalous points varies with time.

All observations started and ended in one and the same control station to check on the operation of emanometers and adjust emanation intensity values when using different ionization chambers.

Magnetic and seismic prospecting

Magnetic prospecting was part of geophysical investigations to ensure reliable mapping, identification and tracing of fault zones. For magnetic prospecting use was made of the same profiles as for electric profiling and emanation survey. These profiles usually intersected the fault zones identified on the basis of geological and geomorphological evidence.

The M27 magnetometer enabled the vertical component of the magnetic field to be measured. The instrument scale division is 10 gammas. Observations started and ended at a permanent control station. Temporary field stations were also used in the process. The point spacing varied from 1 to 10 m.

Combined analysis of the graphs ΔZ and P_k and those showing variations in radioactive emanation intensity enable the mineralogical composition and disturbances of bedrocks to be judged, especially in exposed areas.

Special emphasis was placed on the geophysical verification of faults already identified in the course of geological survey.

Use was made of combined and symmetric electric profiling with a point spacing along the profile of 1–10 m, continuous seismic profiling with an impact method for generating elastic vibrations and a geophone spacing of 1 and 2 m, emanation survey with a 2.5- and 10 metre spacing of air sampling points, and magnetic prospecting. In cases where large drift thicknesses were involved, data on vertical electric sounding were employed to judge fault zones in rocks.

In measured geophysical fields there is some evidence of fault, fractured and crush zones occurring in the near-surface layer.

The combined and symmetric electric profiling graphs feature p_k maxima. The physical nature of this phenomenon is that with air filled interstices, the electrical conductivity in zones of disturbance is lower than that in undisturbed rocks.

In slightly fissured rocks (granodiorite, diorite, andesite, porphyrite), electrical resistivity ranges from 450 to 1700 ohm·m, whereas in highly fissured and kaolinized rocks this figure drops to 40–400 ohm·m.

Diffracted waves are traced on seismic records. The physical nature of this phenomenon is that fracture walls present obstacles to propagation of direct and surface waves which diffract on tension fractures or other vertical and inclined boundaries, on 'diffraction elements' which can be considered as a new source of vibrations propagating in all directions, including the one opposite to the direction of waves generated by the actual source (impact). That is why coherent line-ups of diffracted waves are arranged on seismic records at an angle to those of direct, refracted and surface waves. As evidenced by the analysis of the wave pattern on seismic records, the most clearly defined are diffracted waves caused by a high-energy train of surface waves incident upon the tension fracture or some other vertical boundary with fractures coming out to the surface or under drift whose thickness is less than the length of Rayleigh waves

(Mindel and Golubkov, 1974). Lower intensity of vibrations once waves pass zones of disturbance in a rock mass and interference of waves of different types are additional indicators on seismic records which help identify faults.

A decrease in elastic wave velocities in crush and fractured zones is determined in cases where these zones occurring in rocks of homogeneous lithology are a few tens of metres in size. Should their width be minor, velocity variations are not substantial, and their presence cannot be detected with a given observation system.

The active zones of disturbance feature an increased content of thorium and radium emanations. Unstable geodynamic state rocks in such zones, concentration and variations of stresses, minor movements and vibrations in zones of disturbance caused by small earthquakes, microseisms, are all responsible for 'squeezing' additional emanation portions out of rock interstices. It might be interesting to note that emanation survey is in fact the only geophysical method which enables one to judge the dynamics of the phenomenon. However, emanation survey data are sometimes ambiguous. No emanation anomalies were recorded seven years later over seismic dislocations formed in 1967. All the above-mentioned electrical prospecting and seismic exploration methods have made it possible to judge the presence or absence of zones of disturbance, including those associated with fresh seismic dislocations.

Figure 7.3 illustrates an example of studying a fault zone in the area of the Erdenet Mining and Concentration Complex using emanation survey

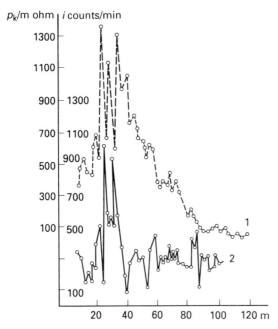

Figure 7.3 Identification of tectonic faults based on results of combined electric profiling (1) and emanation survey (2)

and combined electric profiling. As is evident from the diagram, the emanation intensity within the zone is 50–900 counts/min with a background of 300 counts/min, and p_k values exceed 600–1000 ohm·m with an array length $AO = OB = 15$ m. Based on emanation survey and electric profiling the width of the fault zone is 30–60 m.

Two trial trenches were driven to ascertain the nature of the recorded anomalies and to observe the fault zone across its whole width, one 52 m long, in the area featuring anomalies recorded on the basis of all geophysical methods, and another 70 m long, in the area featuring anomalous electrical resistance and anomalous symptoms in the wave pattern on seismic impact records, yet without the emanation anomaly.

The results of detailed (1 or 2 metre point spacing) measurements along trench 1 and the composition of rocks exposed on the trench wall and floor have shown that the emanation anomaly recorded as a single peak at 5 m spacing is divided into individual peaks (double-humped curve), each peak being associated with weathered granodiorite. Under the impact of a hammer these granodiorites disintegrate and turn into gruss. High values of electrical resistance on the p_k graphs are also associated with the disintegrated granodiorite areas. Minima on the emanation intensity graphs and p_k are associated with porphyrite and andesite dykes which are in fact fissured, yet fairly strong, as compared with granodiorite rocks.

Thus, the results of geophysical investigations in combination with trial trenches helped to trace the tectonic crush zone reliably.

7.2   Other indirect methods

Carbon dioxide survey

The presence of carbon dioxide in the soil atmosphere may also be regarded as an indicator of fracturing in underlying rocks. High contents of carbon dioxide over fractured zones and minor faults were detected by Yegorov (1981) in Moldavia in areas composed of Sarmatian deposits, namely compact clay and sand.

The engineering–geological survey in combination with drilling, mining and an emanation survey made it possible to identify tectonic fractures and faults accompanied by feather jointing in clays and cover loams. Maximum fracture opening is 0.1 cm. Clay porosity is between 2 and 12%, which enables groundwater and gas to circulate along the fractures. Many fractures are revived by the landslide process, yet their origin cannot be traced to landslides since they occur outside the landslide zones. Yegorov formulated major factors pointing to the tectonic origin of fractures, namely regular arrangement, considerable length, and persistence of fracture parameters both horizontally and vertically. Fractures are apparently revived during earthquakes of maximum magnitude 7, which are fairly common in this area.

The CO_2 survey was conducted to trace faults in the landslide slope where they had been previously detected by emanation survey. Measurements were taken with a Riken gas indicator (Japan) with a 0.01% scale division. The total content of carbon dioxide and methane was determined as well as the content of methane once carbon dioxide had been removed with the aid of a hydrochemical absorber. The final content of carbon dioxide was determined on the basis of the difference of the two measurements. A sampling depth within the range 0.2–0.7 m was found not to affect the carbon dioxide content. In the process of air sampling the content of carbon dioxide varied (Figure 7.4). It increased during the first minute and reached a maximum depending on the atmospheric pressure and structural features of the soil. During the second minute the carbon dioxide content dropped almost to zero in areas with no fractures. The carbon dioxide content remained stable within a distance of 0.15 m over faults.

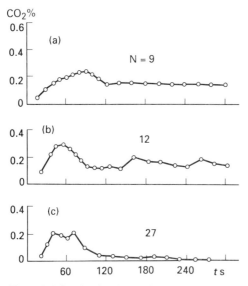

Figure 7.4 Graphs showing variations in concentration of carbon dioxide in soil atmosphere samples: N – number of measurements; (a) over longitudinal faults; (b) over transverse faults; (c) outside fault zones (after Yegorov, 1981)

Thus, carbon dioxide content is an indicator of permeability of rocks overlain by a thick sequence of drifts. Carbon dioxide survey can be employed only in combination with other methods. This method calls for verification and may prove effective in searching for and tracing active faults.

Helium survey aimed at detecting helium content in the soil atmosphere is conducted using similar methods.

Flow rate and resistivity metering

Water flow rate metering is performed directly in a hole. In fractured rocks it is used to identify hole sections with high opening of fractures into which water is absorbed or out of which it flows. The solution of the problem is based on the fact that flow rate of water along the hole axis is measured only within intervals composed of permeable rocks, whereas in confining beds it is constant or equal to zero. Flow rate variation with depth (Figure 7.5) is a step function. Depth, thickness and water-holding capacity of individual fracture zones may be determined. The boundaries of layers differing in permeability properties are identified by points of inflection of the curve.

A flow rate is measured with the aid of a flowmeter whose main component, an impeller, is rotated by water (Grinbaum, 1975). The instrument is moved along the hole from the bottom to the dynamic or static groundwater level. In the event of low flow rates in non-flowing wells, use can be made of a resistivity meter to measure water resistivity. For this purpose water in the hole is salinized with sodium chloride to reach a concentration of 2–4 g/l.

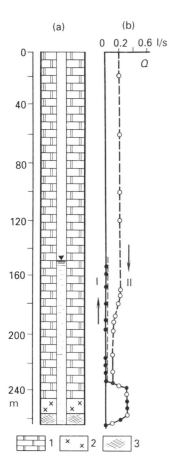

Figure 7.5 Flow rate metering data: (a) lithological column; (b) flow rate diagram at static level (I) and under filling (II); 1 – dolomite; 2 – phosphorite; 3 – crush zone (after Grinbaum, 1975)

A case in point (Figure 7.5) is a well core-drilled in the vicinity of the town of Dzhezkazgan (Kazakhstan). The well was drilled in Ordovician dolomites to a depth of 265 m. The static groundwater level was at a depth of 153 m. Highly fractured dolomites occurred at the bottom of the uncased well. The flow rate diagram I (Figure 7.5(b)) shows water inflow over a depth range of 233–239 m at a rate of 0.3 l/s and outflow over a depth range of 260–265 m. Then the hole was filled with water at a constant rate of 0.2 l/s. As is evident from the flow rate diagram II, some amounts of water were absorbed at a depth of about 180 m, thus pointing to fracture opening. The flow rate is constant until a depth of 233 m is reached. So it may be presumed that there are no open fractures in this interval. In a depth range of 233–239 m the flow rate increases due to the water inflow out of the hole walls. This interval was found to contain highly fractured dolomites. No closeness of fractures was detected in drilling; therefore, it may be suggested that in this interval fractures are only opened. The flow rate metering techniques, just as most indirect methods, fail to record variations of individual fracture parameters. Over a depth range of 260–265 m the hole walls absorbed all injected and inflowing water.

Hence, flow rate metering enables identification of open fracture zones in a rock mass.

Hydrogeochemical and piezometric observations

The use of groundwater flows as indicators of zones of increased fracturing and permeability is possible on the basis of geological survey data and groundwater level and composition observations. However, these possibilities are not taken advantage of, with more costly and technically sophisticated methods being employed on an increasingly large scale.

The presence of springs, groundwater seepage, surface heating or cooling by groundwater, all these factors may serve as fracture zone indicators. They are well interpreted on aerial photographs and during aerovisual observations. Hill top swamping and perched water suggest low permeability and hence indicate that a rock mass is not heavily fractured.

Investigations of the Ust-Ilim foundation revealed a perched water lens occurring on the left bank of the Angara at an elevation of 100 m all the year round. This fact pointed to the presence of water-confining traps in the area. Injections proved that there were no open fractures in the rock mass with perched water at the top. On the opposite bank of the Angara the groundwater level coincided with the river water level in almost all the holes. The river water level fluctuations were instantaneously transmitted for hundreds and thousands of metres, a fact indicative of a large number of open fractures. Subsequent water injections confirmed the presence of open fractures in the area. The groundwater regime was the first indicator of open fractures.

Hydrochemical studies of seawater compositions (near-bottom layers) make it possible to detect large fault zones in the sea floor. Minor faults are detected on the basis of anomalous variation of the chemistry of silty water in the upper layer of bottom sediments (Batoyan and Korolyov, 1976).

Injection of water and air into holes

Injection of water and air into holes for indirect assessment of permeability and fracturing is employed in construction of hydraulic projects and engineering–geological surveying. The method is based on the relationship between water and air absorption by a rock mass and fracturing.

In the event of water injection an indicator of fracturing is specific water absorption, a flow rate per 1 m of hole and 10 m of head. A flow rate of 1 l per 1 m of hole and 1 MPa pressure is taken as a unit of measurement referred to as the 'lugeon' named after French engineer M. Lugeon who pioneered this technique about 60 years ago.

For a water-saturated rock mass:

$$q = Q/hl$$

where

q is specific water absorption, lugeon;
h is head (pressure) over static groundwater level, MPa;
l is length of absorbing hole section, m;
Q is injected water flow rate, l/min.

For a dry rock mass specific water absorption is calculated using the following formula:

$$q = 14.3 \ Q/l^2$$

(Kolichko and Chernyshev, 1972).

It is impossible to pass from specific water absorption to permeability and conductivity used in hydrogeological calculations. That is why specific water absorption should be considered as a fracturing indicator.

Specific water absorption is a relative characteristic. Therefore, a standard method (Karpshev, 1978) is recommended in this respect. It differs from techniques elaborated by western geologists in using a relatively low water pressure, thus ruling out both hydraulic fracturing of formations and suffosion.

Absorption of water as it is injected is largely determined by fracture width, spacing and arrangement relative to the hole.

As regards specific water absorption, rock masses may be grouped as follows:

- practically impermeable, q less than 0.1 lugeon;
- slightly permeable, q ranging from 0.1 to 1 lugeon;
- permeable, q ranging from 1 to 10 lugeon;
- highly permeable, q ranging from 10 to 100 lugeon;
- very highly permeable, q more than 100 lugeon.

Actually there are no rock masses which can be wholly attributed to any of the above-mentioned groups. Specific water absorption in a rock mass varies over wide limits. Hence injections provide an effective means for fracturing evaluation. Variations in specific water absorption by a few orders actually take place not only in a rock mass as a whole, but also in an individual hole. These variations are generally random in nature. Identification of permeable or fractured zones is possible only through statistical analysis of injection results as random values. In this respect injections do not differ from other indirect methods for determining rock mass fracturing.

Injection of air into holes is used in indirect assessment of fracturing in dry rocks. A modification of the method for injecting air into boreholes from openings, pits, and outcrop surfaces has been developed in the USSR at the Leningrad branch of the Gidroproyekt Research Institute (Sorokin, Badukhin and Shpakovsky, 1972). A rate of pressure drop of compressed air in a standard cylinder connected to the borehole section under study is taken as a fracturing indicator. The air permeability factor equal to pressure drop in a cylinder per borehole unit length and unit time is a fracturing characteristic. The unit of measurement is referred to as VOT (Russian abbreviation for Air Fracture Testing): 1 vot = 1 atm/min·m. The VOT method is used in determining the thickness of a zone of weathering, tracing zones of increased fracturing at faults, and identifying rock mass anisotropy due to fracturing. Correlation between a VOT factor and such characteristics of a rock mass as modulus of deformation and porosity is being determined.

The VOT method provides an effective means for studying fracture irregularities. It can be used under diverse conditions with the exception of rock masses with wide open fractures or in flooded strata. The VOT factor ranges from less than 2 vot in slightly fractured rocks to 3–12 vot in zones of increased fracturing to over 12 vot in zones with open fractures produced by release and weathering.

Part III
Solutions of engineering problems based on data on fracture patterns

Chapter 8
Location of folds and faults based on distribution of tectonic fractures

8.1 The belt method

The spatial position of faults and folds is usually determined by methods of structural geology which do not call for fracture patterns to be studied. In this case remote sensing techniques are generally employed. Sometimes, it is impossible to apply standard methods because of poor exposure or absence of exploration holes. Then use can be made of fracture set parameters to determine the position of a fault plane or fold axis, the direction and amount of displacement of limbs, and so on.

The belt method is used in studying tectonic faults. It is based on data on fault-line fractures discussed in Chapter 2. Information on feather joints integrated with other data on faults obtained from surveys can be used to determine (i) the spatial position of a fault; (ii) the direction of displacement; and (iii) apparent displacement and amount of displacement.

The position, dip and strike of a fault plane sometimes cannot be determined during survey. However, the problem can be solved by studying fault-line fractures.

To determine the spatial position of a fault plane, the positions of at least two straight lines in this plane have to be known. One such line is a line of intersection of a feather joint with the fault plane (Figure 8.1). With a large number of feather joints forming a belt on the circular diagram, this line is defined as a belt axis. The other line may be assumed to be the one connecting any two points, A and B, wherein the presence of the fault is certain. In this case to determine fault dip and strike, the inclination of the line AB and its azimuth in the direction of rise have to be measured. Such a direction of the line AB is taken if the upper hemisphere projection is employed for plotting circular diagrams. In cases where lower hemisphere projections are involved (Adler *et al.*, 1961), azimuth of the line AB is taken in the direction of plunge. In addition to measuring parameters of the line AB, large-scale measurements of feather joints are required. Experience has shown that a total of some 1000 joints, or more, at a number of points have to be measured. Desk studies involve plotting a general circular diagram of fracturing (Figures 8.2 and 8.3) with identification of a feather belt or its symmetry axis. Plotted on the same diagram as a point is the line AB (its azimuth is taken circularly and angle, radially from the circumference to the centre). An arc of the meridian swung from A to B represents a fault plane, whose dip and strike can be easily measured on the same diagram.

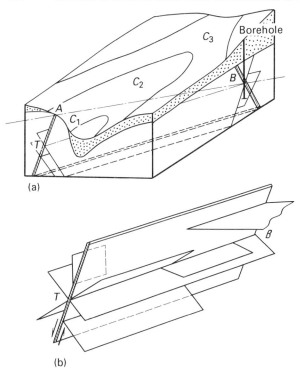

(a)

(b)

Figure 8.1 Concealed fault and its feathering: A – single outcrop wherein the fault is traced; B – hole intersecting the fault; BT – direction of lines of intersection of the fault and feather joints; C_1, C_2, C_3 – outcrops with feather joints

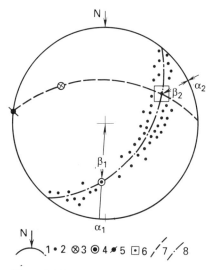

Figure 8.2 Determination of position of tectonic fault plane and direction of displacement by the belt method: 1 – azimuth reference point (north); 2 – fracture pole; 3 – fault belt pole; 4 – fault surfaces; 5 – trace of fault on the ground surface AB as a point; 6 – direction of displacement; 7 – fault surface as a line; 8 – fault feather belt

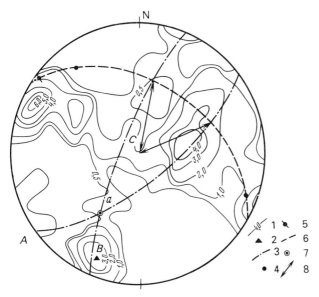

Figure 8.3 Determination of position of fault plane and direction of displacement for one of the areas of the Daregor mining field, Iran (600 measurements of fracture dip angle and azimuth): 1 – point density isolines are expressed on a percentage basis; 2 – dip and strike of beds; 3 – feather belts; 4 – feather belt poles; 5 – azimuth and angle of line *AB* (see Figure 8.1) determined in two fault outcrops; 6 – fault plane; 7 – fault plane pole; 8 – direction of displacement

The method was used in determining the position of 16 faults during the geological exploration of the Kerman coal-field (Iran). Later, five of these faults were intersected by exploratory openings which enabled their dip and strike to be determined to a high degree of accuracy. Thus, the belt method was tested experimentally to determine fault dip and strike (Table 8.1).

Table 8.1 Belt method testing

Fault dip and strike (degree)			
belt method based on fracture studies		*direct measurements in openings*	
dip azimuth	*dip angle*	*dip azimuth*	*dip angle*
320	66	316	56
292	64	295	68
330	46	336	36
140	58	124	43
113	72	110	75

The second problem, determination of displacement direction, has been solved previously (Budko, 1958). Direction of displacement is known to coincide with the line of intersection of the feather belt and fault planes (see Figure 8.2).

The third problem, determination of apparent displacement/amount of displacement, can be easily solved geometrically provided the first two problems have been solved. Apparent displacement

$$a_c = a'_c / \cos \varphi$$

where

a'_c is measured apparent displacement;
φ is angle between direction along which a'_c is measured and that of limb displacement (Figure 8.4)

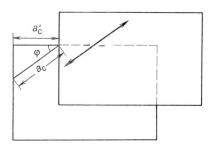

Figure 8.4 Angle between direction of displacement and direction of measuring amount of displacement on the fault plane

8.2   Method of revolution

This method is used to study structural fold patterns. It is based on general rules governing the distribution of fold fractures described in Chapter 2.

The method makes it possible (i) to check whether a system of fractures belongs to fold fractures or to fractures formed prior to folding; (ii) to show that a fracture system originated after folding; (iii) to determine the position of a fold axis based on the arrangement of fracture systems in cases where only limbs are exposed or only one limb is accessible to observation.

The method involves conversion of the angular coordinates of the circular diagram representing dip and strike of the fold limb and related fractures. Conversion is accomplished by rotating the circular diagram; hence the name of the method. The previous modifications of the method involved double conversion of coordinates, which is not correct. Triple conversion to fit the three Cartesian axes is imperative.

A system of fractures in a fold limb is known to follow a particular pattern. It comprises five major sets, two parallel and one perpendicular to the fold axis, and the remaining two are diagonal, yet perpendicular to the layer. Each of the sets is oriented relative to the plane of the layer and the axis of the fold parallel to the layer plane. Take the fold axis position as the X-axis and the perpendicular to the layer as the Z-axis, the Y-axis is arranged in the layer plane along the line perpendicular to the X-axis. It should be noted that it is only in specific, although fairly common, cases that the X-axis coincides with the layer strike, and Y-axis with the layer dip. The position of the fold fracture sets relative to the aforesaid

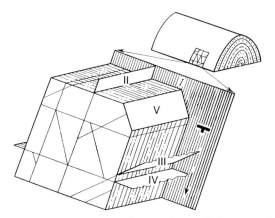

Figure 8.5 Position of major sets of tectonic fractures relative to bedding surface and fold axis: I–V – system numbers

coordinate axes is fixed (Figure 8.5). However, the coordinate set features particular orientation at each point of the fold. For example, the Z-axis is directed towards the zenith on the fold axis, provided that the fold axis is horizontal and inclined in different directions in the limbs. Using the method of revolution the coordinate systems are converted until the axes are fully aligned with each other (three operations). Should the fracture sets coincide once the axes are aligned, this suggests that they were formed prior to folding or are associated genetically with the fold (Figure 8.5). If the fracture systems fail to coincide, presumably they were formed after folding.

Coordinate conversion can be accomplished for each particular point graphically on a circular diagram. The circular diagram plotted on some transparent material is superimposed on a spherical coordinate grid. First, the X-axes of all the diagrams are aligned and for this the fold axis is brought to the horizontal position in each diagram.

The point of intersection of the fold axis with the coordinate sphere (it is similar to point A in Figure 6.2) is made coincident with the northern radius and transferred to the north pole of the grid, which means actually that the entire coordinate system XYZ turns through the angle equal to fold axis inclination.

Second, the Y-axes are aligned. For this the layers from different blocks are referenced to a single dip azimuth. The bedding pole is brought to the northern radius by rotating through angle α. All the points shown in the diagram are rotated through the same angle and in the same direction.

Third, the Z-axes are aligned. For this the bedding pole is transferred to the centre of the diagram. As for other points in the diagram, they are shifted along the meridians through the same angle in the southern direction.

Now the Z-axis is directed toward the zenith, the Y-axis northward, and the X-axis eastward. With the rotation procedure performed three times, we can have either of two situations:

1. All the fracture systems coincide with transparent circular diagrams superimposed on each other. This suggests that they were formed prior to or concurrently with folding.
2. Some fracture systems coincide and some do not, which indicates that the latter were formed after folding.

As an example, the method of revolution was used in studying fractures in the Kerman coal-field. To ascertain the origin of fractures in the coal-bearing strata, a general diagram of fracturing (Figure 8.6(a)) was plotted with data on six tectonic blocks located on different elements of large-folds spaced 20–60 km apart in the area. A total of 500–10 000 fractures was measured in each block. Over 15 000 measurements were taken to plot the diagram. Such a large number of measurements made it possible to obtain reliable information. The sequence of operations was as follows:

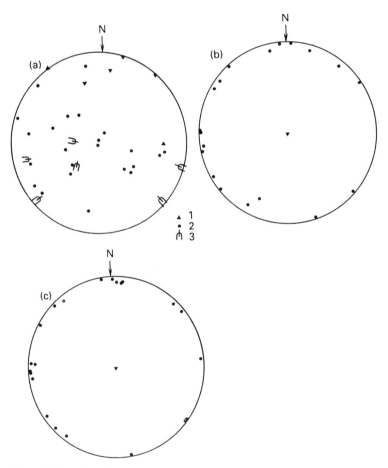

Figure 8.6 Locating layers with reference to the horizontal plane: (a) actual position of fracture system; (b) position of systems once located with reference to horizontal plane with fold axis position ignored; (c) same with fold axis position taken into account; 1 – dip and strike of beds; 2 – dip and strike of fracture set centres; 3 – fold axis position

(i) locating the fold axis with reference to the horizontal plane; (ii) locating layers from different blocks with reference to a single dip azimuth; (iii) locating layers with reference to the horizontal plane. Without the first operation (previously omitted), the other two failed to make fold systems in Figure 8.6(b) coincident. A statistical check has shown that the circular distribution of points in Figure 8.6(b) does not differ from uniform. Following the three operations (Figure 8.6(c)) the hypothesis on uniform distribution is rejected. There are reasons to believe that the points are not grouped randomly. Hence, the fracture systems are arranged identically relative to the stress field axes. Logically, these systems can be associated with the stress field responsible for the fold since their arrangement relative to the fold elements (limbs, axis) is not only identical in different portions of the fold but also agrees with the position of the stress axes known from the theory of strength.

Thus, location with reference to the horizontal plane is to be accomplished in three stages in accordance with tectonophysics theory.

The position of a fold axis may be determined from the arrangement of fold fractures.

If a fold axis is not exposed and only one fold limb is exposed, the fold axis position is obscure. In this case to determine the position of the fold axis, reliance has to be placed on fracture data. This can be understood by seeing how some specific problems are solved.

Problem 1: determination of the fold axis position by measuring dip and strike of bedding in two limbs

The problem is solved in a circular diagram with the bedding poles plotted (Figure 8.7(a)). A meridian is drawn through these poles. The fold axis is perpendicular to the meridian plane. Dip and dip direction of the fold axis

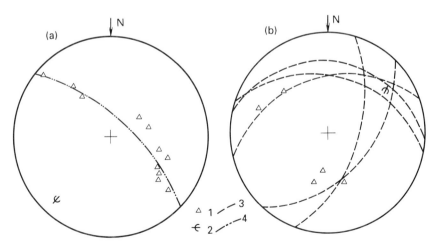

Figure 8.7 Determination of azimuth and inclination of fold axis on the basis of bedding dip and strike: (a) method I; (b) method II; 1 – bedding set pole; 2 – fold axis rise direction; 3 – bedding surface as a line; 4 – surface perpendicular to fold axis

are unambiguously determined in the diagram. The same problem can be solved in a different manner (Figure 8.7(b)). Plotted on the circular diagram are meridians representing bedding planes at observation points. With the meridians representing different limbs of the fold intersecting, the fold axis position becomes evident. In this case the most probable position is taken as an average of the meridian intersection points.

As examples, consider problems solved by Chernyshev and Pogrebisky during their investigations in Tajikistan. The solution with the first method was obtained for the southern portion of the Fark–Dzhurudzh syncline with its rising axis. There the syncline is composed of Lower Cretaceous red beds and Upper Cretaceous carbonate deposits. Measurements of bedding orientation were taken at 11 stations. Station numbers 140–142 gave the following dip and strike characteristics in the north-western limbs: $321°\angle 54°$, $392°\angle 41°$, and $310°\angle 82°$. Stations numbers 123, 125, 127, 143, 146–149 gave the following dip and strike characteristics in the south-eastern limb: $128°\angle 54°$, $121°\angle 48°$, $129°\angle 68°$, $56°\angle 30°$, $108°\angle 53°$, $102°\angle 42°$, $121°\angle 40°$, and $76°\angle 34°$. Determination of the fold axis inclination using the first method gave a rise azimuth of 217° and an angle of 10°. The second method provided the following data: rise azimuth 217°, rise angle 15°. The second method (see Figure 8.5(b)) is exemplified by studies in Tajikistan in the area where the Dzangou anticline abruptly plunges south-westwards. Its limbs are composed of lower Palaeogene limestones. Five stations were used to take measurements. Station numbers 106 and 108 gave the following dip and strike characteristics in the north-western limb of the anticline: $285°\angle 63°$ and $310°\angle 50°$ respectively. Stations numbers 102, 109 and 114 gave the following dip and strike characteristics in the south-eastern limb: $190°\angle 43°$, $180°\angle 32°$, and $158°\angle 45°$. The graphic solution of the problem is illustrated in Figure 8.5(b). The result is as follows: the Dzangou anticline axis plunges south-westwards (azimuth 277°, angle 29°). From the evidence of the geological map, the fold plunge angle in the area is about 40°.

Problem 2: determination of the fold axis position based on the orientation of fold fractures where only one fold limb is accessible

The solution is based on the relationship (see Chapter 2) between the orientation of fold fractures and that of the fold axis. Of the four fold fracture sets (Figure 8.5), two tension fracture sets are assigned even numbers and two shear fracture sets are assigned odd numbers. The tension fracture set II is perpendicular to the axis. The tension fracture set IV is parallel to the axis. It is possible to single out two major tension fracture sets using their morphological characteristics. It is much more difficult to distinguish between longitudinal (IV) and transverse (II) fracture sets relative to the fold. Therefore, based on tension fractures, one can obtain two solutions of the fold axis position problem, differing from one another by about 90°. The solution can be verified on the basis of the orientation of shear fractures (sets III and V). The fold axis forms angles of 0 and 90° with bisectrixes of angles between these sets. The acute bisectrix generally differs from the fold axis position by 90°, whereas the obtuse bisectrix coincides with the fold axis direction.

As an example, consider the orientation of the fold axis determined on the basis of fracturing data from the north-western limb of the Fark-Dzhurudzh syncline. Orientation of 600 fractures was measured at three stations. There are five sets in the general diagram (Figure 8.8) whose numbers match the designations in Figure 8.5. The axial direction is close to point II or point IV. Since the area is known to have no folds with axes close to vertical, the solution with point IV is to be discarded. As regards point II, the fold axis position is verified as follows: (i) coinciding with point II; (ii) differing from point IV by 90°; (iii) differing from the bisectrix between points III and V by 90°. The verification gives a fold axis azimuth of 229° and an angle of 10°. This result agrees well with one previously obtained by another method (227° < 10°).

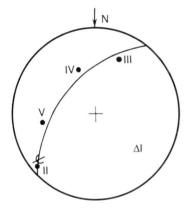

Figure 8.8 Determination of azimuth and inclination of a fold axis on the basis of the dip and strike of tectonic fractures

Chapter 9
Fractures and faults as indicators of previous earthquake parameters

9.1 Multivariate regression relationships between maximun seismic dislocation and earthquake magnitude and intensity

Faults and earthquakes are interrelated both in space and time. Fault parameters such as length, width, amount of displacement and width of gaping fissure are used as indicators of earthquake intensity in scales of seismic intensity, in particular scale MSK–64. Details of the Medvedev–Sponheur–Karnick 1964 intensity scale (MSK–64) are set out in sufficient detail in an appendix to Skipp and Ambraseys (1987, 18/21–18/23). The same authors reviewed the characteristics of earthquakes, assessment of earthquake hazard, and the ground effects caused by earthquakes. Fault parameters were used by Solonenko (1970) as major features for drawing up a special scale and by Belyi (1974) for classifying residual seismic deformations and seismic displacements.

The statistical relationship between parameters of seismic faults and those of earthquakes has been analysed in the following papers (Tocher, 1958; Iida, 1959; Dedova, 1967; Rats and Chernyshev, 1970; Solonenko, 1973). However, all those papers were generally based on limited data and failed to make the most of potentialities offered by statistical methods. Here an attempt is made to make use of the advantages of multivariate statistical analysis successfully employed in engineering geology.

Multivariate analysis helps to reconstruct earthquake parameters to a high degree of accuracy. Information on a total of 100 earthquakes which occurred in different parts of the world was used for the statistical analysis.

The multivariate relationships between earthquake intensity and seismic dislocation parameters were used in detailed seismic zoning of the area of the Erdenet Mining and Concentration Complex in Mongolia.

Relationship between the parameters concerned was first ascertained quantitatively by Tocher (1958) who obtained the following equations on the basis of data on 10–12 earthquakes:

$$M = 5.65 + 0.98 \log L$$
$$M = 5.22 + 0.53 \log AL$$

where

M is earthquake magnitude;
L is seismic dislocation length, km;
A is amount of displacement, m.

A similar study, though based on ample evidence, was made by Iida (1959) who obtained the following equations for the Japanese islands (left) and worldwide (right):

$M = 6.6 + 0.5 \log L$ $\qquad\qquad$ $M = 6.3 + 0.6 \log L$

$M = 6.2 + 0.4 \log AL$ $\qquad\quad$ $M = 5.4 + 0.5 \log AL$

$\log E = 21.7 + 0.7 \log L$ $\qquad$ $\log E = 21.2 + 0.9 \log L$

$\log E = 21.1 + 0.5 \log AL$ $\qquad$ $\log E = 19.9 + 0.7 \log AL$

where E is earthquake energy, erg.

According to Solonenko (1973), $\log L = (1.01 \pm 0.02)M - 6.18$. He also obtained relationships between the area of residual seismic deformations (S) and earthquake magnitude in the form of:

$$\log S = (0.99 \pm 0.07)M - 3.6$$

Orthogonal regression was used for this purpose by Rats and Chernyshev (1970). The following equations were obtained:

$$\log E = 1.86 \log A + 22.18$$
$$\log E = 1.80 \log L + 14.24$$

where E is earthquake energy, erg.

Relationship between earthquake intensity I_0 and width B of seismic fractures was studied by Dedova (1967), who introduced a new earthquake parameter (intensity) and a new seismic dislocation parameter (width). Based on the seismic microzoning pattern and considering seismic dislocations secondary to faults, different types of ground were distinguished.

Multivariate regression analysis was applied to the problem of studying a relationship between earthquake and seismic dislocation parameters (Chernyshev, 1974; Anon., 1975). Based on data on 75 earthquakes, the regression equations relating earthquake intensity I_0 and $\log L$, $\log A$, $\log B$ were obtained. Use was made of the REGR regression analysis program developed by Petrov and Ellansky based on simple regression algorithm and approximating sought relationships with a polynominal of degree n. Polynomials of degrees 1, 2 and 3 were calculated. Then polynomial members of low informativeness were withdrawn. The equations were checked against information not used when deriving them. As a result (Tables 9.1 and 9.2) an increase in the polynomial degree above the first did not prove to substantially reduce a forecast error. Therefore, linear equations were used subsequently in approximation.

The equations yielding minimum errors once checked are given below

$$M = 5.53 + 0.18 \log L + 0.64 \log A, \; \hat{s} = 0.46$$

where $\hat{s}$ − standard forecast error

$M = 5.82 + 0.04 \log L + 0.07 \log A + 0.03 \log^2 L + 0.10 \log L \log A + 0.14 \log^2 A$
$\hat{s} = 0.41$
$M = 5.30 + 0.11 \log L + 0.76 \log A + 0.17 \log B, \; \hat{s} = 0.39$
$M = 5.35 + 0.11 \log L + 0.33 \log A + 0.15 \log B + 0.05 \log^2 L$
$\qquad + 0.75 \log A \log L + 2.03 \log A \log B - 0.78 \log L \log B + 1.02 \log^2 L$
$\qquad - 0.86 \log^2 B, \; \hat{s} = 0.27$

Table 9.1 Results of verifying relations obtained with REGR program

Tabular intensity based on literary data	Deviation of calculated values from tabulated values: $I_{o\,cal} - I_{o\,tab}$															
	$I_o = f(\log L, \log A, \log B)$ degree of equation				$I_o = f(\log L, \log A)$ degree of equation				$I_o = f(\log L, \log B)$ degree of equation				$I_o = f(\log A, \log B)$ degree of equation			
	1	2	3'	3	1	2	3'	3	1	2	3'	3	1	2	3'	3
10.0	0.6	–	0.8	0.6	0.5	0.7	0.5	0.6	0.6	0.8	0.9	0.7	0.7	0.8	0.9	0.9
11.0	-0.7	–	-2.8	-1.1	-1.2	-1.4	-1.5	-1.4	-0.6	-0.5	-0.5	-0.6	-1.1	-1.2	-0.7	-1.0
9.0	0.9	–	0.9	0.5	0.6	0.4	0.2	0.3	1.0	0.9	0.8	0.7	0.7	0.4	0.6	0.6
9.5					-0.8	-1.1	-1.3	-1.2								
10.5					-0.6	-0.8	-0.9	-0.8								
11.0					-1.0	-1.0	-1.2	-1.1								
11.0					-0.7	-0.7	-0.7	-0.6								
10.0					0.2	-0.1	0.1	0.2								
8.0					0.6	0.2	0.0	0.1								
11.0					-1.5	-1.9	-1.8	-1.8								
9.0					-1.5	-0.9	-0.7	-0.8								
9.0					0.8	0.6	0.4	0.5								
10.0					0.1	0.1	-0.1	0.1								

Note: 3' is reduced polynomial of degree 3.

Table 9.2 Additional results of verifying equation $I_0 = f(\log L, \log A)$

Tabular intensity based on literary data	Deviation of calculated values from tabulated values: $I_{0\ cal} - I_{0\ tab}$ degree of equation			
	1	*2*	*3'*	*3*
9.5	−0.8	−1.1	−1.3	−1.2
10.5	−0.6	−0.8	−0.9	−0.8
11.0	−1.0	−1.0	−1.2	−1.1
11.0	−0.7	−0.7	−0.7	−0.6
10.0	0.2	−0.1	0.1	0.2
8.0	0.6	0.2	0.0	0.1
11.0	−1.5	−1.9	−1.8	−1.8
9.0	−0.5	−0.9	−0.7	−0.8
9.0	0.8	0.6	0.4	0.5
10.0	0.1	0.1	−0.1	0.1

$I_0 = 7.25 + 0.30 \log L + 0.92 \log A,\ \hat{s} = 0.62$

$I_0 = 7.49 - 0.47 \log L + 0.69 \log A - 0.12 \log^2 L + 0.57 \log L \log A - 0.07 \log^2 A,\ \hat{s} = 0.51$

$I_0 = 7.98 + 0.52 \log L + 0.24 \log A + 0.34 \log B,\ \hat{s} = 0.54$

$I_0 = 8.10 + 0.04 \log L - 1.26 \log A + 1.50 \log B - 0.06 \log^2 L$
$0.12 \log L \log A + 0.26 \log B + 1.38 \log^2 A - 2.84 \log A \log B$
$\quad + 0.94 \log^2 B,\ \hat{s} = 0.43$

In Anon. (1975) equations relating earthquake parameters to those of seismic dislocations were obtained with the help of orthogonal regression. However, this did not result in basically new empirical relations being obtained.

The relationship between earthquake intensity and seismic dislocation parameters is closer than that between earthquake magnitude and the same parameters. It may be presumed that this stems from the hypocentre depth being unaccounted for. Another feature of the equations is that the relationship becomes closer as the number of equation factors increases. The closest relationship, judging by the value of standard forecast error, $\hat{s}$, is obtained for four-factor equations, then come three-factor and two-factor equations which are not given in analytical form since they are of low practical value. For the above relationships, see Figures 9.1–9.3. A decrease in relationship closeness if the number of factors concerned goes down indicates that all the seismic dislocation parameters under study (amount of displacement, length and width) carry information on earthquake magnitude and intensity. Of those parameters, as a major indication the amount of displacement is most closely related to earthquake magnitude and intensity. Thus, it may be inferred that first of all the maximum amount of displacement has to be determined in paleoseismogeological reconstructions. The earthquake epicentre is apparently associated areally with the zone of maximum displacement. Length and aperture of seismic dislocations come, respectively, second and third (next to amount of displacement) as regards informativeness.

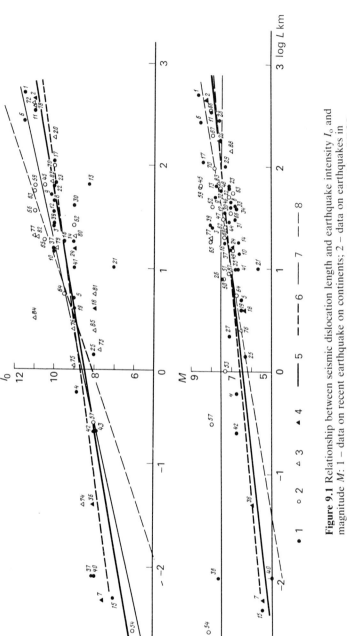

Figure 9.1 Relationship between seismic dislocation length and earthquake intensity I_0 and magnitude M: 1 – data on recent earthquake on continents; 2 – data on earthquakes in littoral areas; 3 – data on old earthquakes; 4 – earthquake data used for checking REGR equations; 5 – orthogonal regression equation graphs based on recent earthquakes on continents; 6 – REGR equation graph based on recent earthquakes on continents; 7 – orthogonal regression equation graph for littoral areas; 8 – orthogonal regression equation graph based on old data. Figures denote earthquake numbers (Anon, 1975)

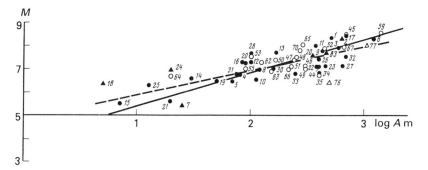

Figure 9.2 Relationship between magnitude, M, and amount of displacement, A, along seismic dislocation. For symbols see Figure 9.1

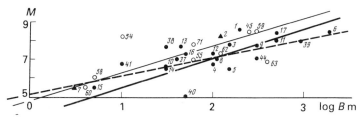

Figure 9.3 Relationship between magnitude, M, and width, B, of seismic dislocation. For symbols see Figure 9.1

9.2 Application of regression equations to the reconstruction of earthquake magnitude and intensity in the area of the Erdenet Mining and Concentration Complex, Mongolia

The Erdenet Mining and Concentration Complex is situated in a seismic area in north-western Mongolia. The engineering–geological survey at the project design stage involved seismic microzoning and detailed seismic zoning to determine the background seismicity of the area.

The territory of the Erdenet Mining and Concentration Complex is situated at the contact of regions with seismicity of intensity 7 and 8. Use was made of a palaeoseismogeological method since only scarce data from instrumental observations were available in the area.

The area is situated within a regenerated epiplatform region. From the north and south it is bounded by the regional Selenga and Bayangol faults trending nearly east-west which are cut by faults trending north-west and nearly north-south (Figure 9.4). Most of the faults were initiated long ago, to be revived through geological time, in particular in the Pleistocene and Holocene. Less than 100 km west of the project area there is a zone of young seismovolcanic faults recognizable through outflows of Quaternary basalts, young volcanic domes, thermal springs and earthquake epicentres, including the 1967 Mogod earthquake of intensity 10. This zone is the western boundary of the area under study. The eastern boundary is a

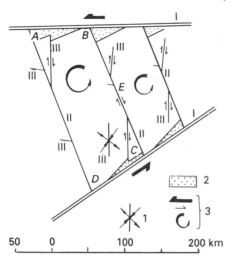

Figure 9.4 Tectonic kinematic diagram of Erdenet environs, Mongolia: 1 – direction of compressive and tensile stresses during the Mogod earthquake; 2 – grabens in tension zone; 3 – block movement direction. Area is ca 200 × 200 km; I, II, III – order of tectonic fault magnitude

depression axis coinciding with meridian 105 which is considered as the boundary between areas of low seismicity in the east and high seismicity in the west. The area of the Erdenet Mining and Concentration Complex situated to the west of this boundary falls within the zone of high seismicity.

The structural pattern of the project area is based on a system of faults of nearly north-south and north-west trends. The area is composed of effusive-sedimentary and intrusive complexes of different ages from the Vendian–Cambrian to recent. Palaeozoic rocks were highly metamorphosed and deformed during orogeny. Mesozoic volcanic sedimentary rocks are folded and faulted. Bedrocks are overlain by a loose Neogene–Quaternary cover varying in thickness.

Geomorphologically, the area is a dissected plateau with elevations of 1000–2040 m situated at the junction of two major ridges in North Mongolia, namely the Khangai and Khentei. It is believed that the relief features are determined by the fact that the area belongs to the post-platform activity belt characterized by smooth movements responsible for vaults and depressions and sharp movements along the disjunctive dislocations of different order and trend. A variety of blocks gave rise to highly dissected topography with horst massifs dominating flat or slightly undulating plains. Faults repeatedly revived by recent movements serve as block boundaries. It is with those faults that most of the earthquake epicentres are associated (Medvedev, 1971).

The most powerful earthquakes have left their imprints on the Earth's surface, namely as seismic dislocations. The largest seismic dislocation was described south of the Mogod-somon. It originated from an earthquake of intensity 10 on 5 January, 1967 in the area of Pleistocene deposits. The energy class was K-16 and magnitude was 7.5. The seismic fault is a zone of fractures extending for 45 km, azimuth 10°. The turf beds and Anthropogenic sand–clay deluvial, proluvial and lacustrine deposits are burst or thrust along the fractures (Figures 9.5 and 9.6). Five years after the earthquake, the vertical walls of open fissures were more than 1 m high. In the tension areas gaping measured a few tens of centimetres. In areas of compression a rampart of turf blocks was formed 1.5 m high and up to 10 m wide. Signs of right-slip with about 2.5 m displacement are evident. Side-by-side with the fresh seismic dislocation in the Mogod intermontane trough, other recent and old seismic dislocations can be seen pointing to the recurrence of earthquakes similar to the one that took place in 1967.

East of the Mogod–somon on the Mogoin–gol river an old seismic dislocation has been described (Figure 9.7). It represents a ravine extend-

Figure 9.5 Scarp of Mogod seismic dislocation (1967) in loams. Vertical feather joints spaced 10–50 cm are visible, the scarp is 0.5 m high

Figure 9.6 Mogod seismic dislocation. 150 metre-long area is visible from the air

Figure 9.7 An old seismic dislocation in Mogoin-gol valley

ing in the right-hand slope of the Mogoin-gol valley, azimuth 350°. The ravine cuts through small watersheds and is turfed. It bears no signs of erosion whatsoever. Its thalweg both plunges and rises. The overall width of the ravine is about 5 m, its depth is 0.3–0.5 m and length is 10–12 km. A feather joint some 150 m long is found in the central portion of the seismic dislocation. The northern termination fades away in the rock mass where parallel gaping fissures about 1 m wide can be seen.

The Mogoin-gol seismic dislocation is about 100 years old. Its appearance is more ancient than that of the strike slip fault which originated from the 1905 Khangai earthquake (North Mongolia) of intensity 11.

Along with the above seismic dislocations, there are numerous landslides, mudflows, collapses, talus deposits and other manifestations of slope instability. These seismogravitational deformations are presumably associated with the effect of gravitational forces and earthquakes.

In addition to the two seismic dislocations, some 30 more similar forms have been described. Table 9.3 lists magnitude and intensity of previous earthquakes determined on the basis of their parameters.

Earthquake epicentres with indication of magnitude and intensity were plotted on the diagram for detailed seismic zoning. This helped to expand the list of local earthquakes and verify the boundary between regions with seismicity of intensity 7 and 8.

9.3   Reconstruction of the position of the axes of principal normal stresses in the earthquake epicentre

Major earthquakes are associated with the instantaneous (in a geological sense) origination of faults extending for tens and hundreds of kilometres

Table 9.3 Earthquake intensity and magnitude calculated from seismic dislocation parameters

| Seismic dislocation No | Seismic dislocation parameters | | | | | | Magnitude | Intensity | Remarks |
| | length (km) | | amplitude (cm) | | width (cm) | | | | |
1	2	3	4	5	6	7	8	9	10
1	45.0	1.65	250	2.4	110	2.0	7.6	10.0	Mogod, 1967
2	10.0	1.0	–	–	100	2.0	7.0	9.5	
3	2.5	0.4	–	–	–	–	7.0	9.0	
4	1.8	0.25	–	–	–	–	6.8	9.0	
5	2.5	0.4	–	–	30	1.48	6.5	9.0	
6	2.0	0.3	–	–	–	–	6.8	9.0	
7	17.0	1.23	–	–	–	–	7.2	9.5	Seismic dislocations 7 and 8 possibly result from earthquake
8	17.5	1.24	–	–	–	–	7.2	9.5	
9	14.0	1.15	–	–	–	–	7.2	9.5	
10	1.0	0.0	–	–	–	–	6.4	9.0	
11	0.8	-0.1	–	–	30	1.48	6.5	9.0	
12	5.0	0.7	–	–	40	1.6	6.7	9.5	
13	5.0	0.7	–	–	50	1.7	7.0	9.0	
14	4.0	0.6	–	–	–	–	7.0	9.0	
15	4.5	0.65	–	–	–	–	6.7	8.5	
16	1.0	0.0	–	–	–	–	6.8	9.0	
17	2.0	0.3	–	–	–	–	6.8	9.0	
18	15.0	1.18	–	–	–	–	7.2	9.5	
19	18.0	1.25	–	–	–	–	7.2	9.5	
20	6.0	0.78	–	–	–	–	7.0	9.5	
21	2.0	0.3	–	–	–	–	6.8	9.0	
22	1.0	0.0	–	–	–	–	6.7	9.0	
23	0.6	-0.22	–	–	–	–	6.6	8.5	Seismic dislocations 23 and 24 possibly result from earthquake
24	0.2	-0.7	50	1.7	15	1.8	5.8	8.0	
25	0.6	-0.22	–	–	–	–	6.6	8.5	Seismic dislocations 25–28 possibly result from earthquake
26	0.15	-0.83	–	–	–	–	6.4	8.0	
27	0.2	-0.7	40	1.6	30	1.48	6.4	8.5	
28	0.2	-0.7	–	–	50	1.7	6.8	8.0	
29	2.0	0.3	–	–	–	–	6.8	9.0	
30	1.0	0.0	–	–	–	–	6.7	8.5	
31	4.5	0.65	–	–	–	–	7.0	9.0	Seismic dislocations 31 and 32 possibly result from earthquake
32	4.0	0.60	–	–	–	–	7.0	9.0	
33	0.6	-0.22	–	–	–	–	6.6	8.5	
34	0.4	-0.40	–	–	–	–	6.6	8.5	

fringed by feather joints. The faults proper have been extensively studied and described in the literature, whereas feathering has received little attention from geologists. Yet feather joints may serve as an indicator of the orientation of principal normal stresses responsible for faults. Reconstruction of the stresses is of importance to the study of the earthquake mechanism.

Let us consider feather joints in the strike slip faults originating from earthquakes in Mogod (Medvedev, 1971), Yanitse (Ketin and Roesli, 1953), and in the Khangai mountains (Voznesensky, 1962). The Mogod fault originated, as mentioned above, from the earthquake that took place on 5 January 1967. In the area covered by detailed studies the eastern limb of the fault was found uplifted by 0.9 m and shifted to the left for a distance of up to 2.5 m. The fault in Yanitse (Turkey) originated from an earthquake of intensity 10 and magnitude 7.5–8.0 that occurred on 18 March, 1953. The fault can be traced for 40 km lengthways. Morphologically, this fault is similar to the one described above. The fault in the Khangai mountains in north-western Mongolia originated from earthquakes of intensity 11 that took place on 9 and 23 July, 1905. Its length is 450 km.

In a zone several dozens of metres wide (in Mogod up to 300 m wide) on both sides of the faults, there are vertical feather joints (Figures 9.8 and 9.9) penetrating into the Quaternary cover beds. Soils are fractured into polygons. Fractures run parallel to the main fault surface or form an acute angle with it. The Mogod fault is characterized by two pairs of systems of Riedel and feather fractures which in pairs form an acute angle of about 68°. Of these systems, two are symmetrical about the main fault which runs along the bisectrix between the fractures (Figure 9.10). The third system stretches along the fault and the fourth forms an angle of about 68° with the fault.

Feathering of the Yanitse fault has similar features. A rose diagram (see Figure 9.10(c)) shows the origin coinciding with the fault trend and located to the left on the horizontal line. Angles between the fractures and the fault trend are plotted in a clockwise direction. The density of fractures in some portions of the rose diagram is shown to scale on the horizontal line. The diagram shows a less pronounced set and two sets of oblique fractures forming angles of 40 and 135° with the fault line. The fault runs to within 5° along the bisectrix of the angle between these sets which is 85°. Besides, there are two sets of fractures (angles 75° and 105°) also symmetrical about the main fault and deviating by 75° from the latter. The feather joints of the seismic faults on the circular diagrams form belts typical of strike slip faults which run along the great circle of the diagrams. Thus, they are similar to the tectonic feather fractures of strike slip faults (Figure 9.11).

The parameters of feather joints feature large standard deviations (Table 9.4). Their width ranges from 0.1 to 1.0 m, spacing from 2.2 to 18.2 m and length from 3.0 to 76 m. Assessment of parameter distribution has shown that the hypothesis on the lognormal distribution of all parameters is not rejected at a high (more than 0.05) significance level. The same result has been obtained from graphical checks of identical hypotheses for fracture width and spacing on the basis of data collected by Voznesensky (1962). With reference to tectonic fractures which are fairly widespread and have

Figure 9.8 Feather joints in the Mogod seismic dislocation, photograph covering an area of 200×300 m²

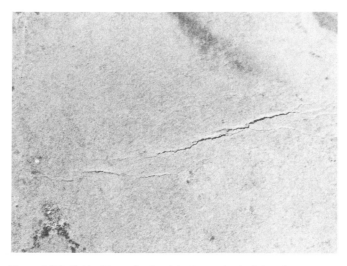

Figure 9.9 Individual large feather joint in a seismic dislocation

been studied most comprehensively, it might be well to point out that the above parameters of these fractures also feature lognormal distribution (Chapter 3). Thus, after comparing tectonic and seismic fractures, it is evident that these are close to each other as regards width, spacing and length distribution.

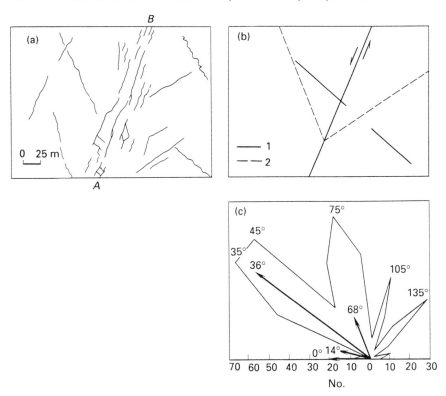

Figure 9.10 Fault-line fractures in seismic dislocation: (a) graphic representation of fractures originating in frozen loams during the Mogod earthquake on 5 January, 1967; (b) arrangement of major fracture sets: 1 – advance, 2 – feather; (c) rose diagram of fault-line fractures in Mogod (arrows) and Yanitse (polygon)

The stress field responsible for feather joints can be reconstructed from the fracture pattern described in a fault-line zone. Using the Mohr criterion of failure, which describes failure of homogeneous isotropic material under rapid loading, rock strength is characterized by an envelope of major principle stress circles. Engineering–geological practice has shown that a rectilinear inclined envelope can be applied to loams. The equation expressing the rock failure condition is in the form of the well-known Coulomb–Navier criterion:

$$\tau = \sigma_n^{\tan} \varphi + c$$

where

τ is limit shear stress in the area of displacement;
σ_n is normal stress in the direction perpendicular to the area;
φ is angle of internal friction;
c is cohesion.

For the loams in question, φ and c determined in surveying the neighbouring area are 22° and 0.25 respectively.

Table 9.4 Checking the hypothesis on laws governing distribution of parameters of seismic dislocation feather joints

Parameters	Number of samples	Average value	Standard devia-tion	number of degrees of freedom	Checking for compliance with the distribution law [*]		
					Pirson criterion		Probability P[**]
					χ^2	$\chi^2_{0.05}$	
Fracture width (cm)	83	28.2	17.2	4/5	84.4/8.9	9.5/11.1	0.301/0.96
Fracture spacing (m)	60	7.4	3.9	3/3	10.7/2.7	7.8/7.8	0.22/0.99
Fracture length (m)	102	20.1	14.5	7/7	23.2/8.0	14.1/14.1	0.501/0.97

Notes
[*] normal – numerator, lognormal – denominator.
[**] Probability of random difference between theoretical and empirical distributions. For example, if $P = 0.301$ the probability of random difference is low, the difference is to be considered non-random, and the hypothesis is to be rejected; if $P = 0.96$, the probability of random difference is fairly high and the hypothesis is not rejected.

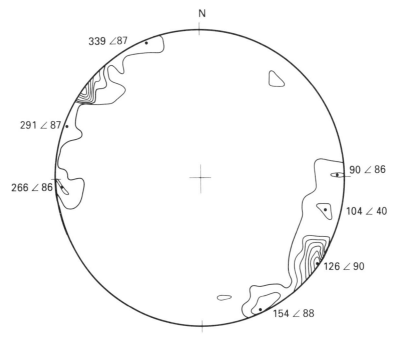

Figure 9.11 Circular diagram of fault-line fractures in Mogod seismic dislocation. Point density isolines are expressed on a percentage (1%) basis

The following proposals can be made with respect to stresses in the fault limbs when feather joints evolve following the initiation of the main fault. The fault is in fact a strike slip fault with marked horizontal, and less pronounced vertical, displacement. The maximum force whose direction is unknown apparently coincided with the direction of maximum residual deformation, i.e. was directed along the line of intersection of the fault with the Earth's surface. Thus the axis of maximum principle normal stresses, σ_1 is in line with this direction. It is hardly probable that some normal stress acted in the direction perpendicular to the fault at the Earth's surface, proof of which is open fractures up to 1 m wide arranged along the main fault zone. Some minor tension might have worked there. So the axis of algebraically minimum principal normal stress, σ_3, is directed along the perpendicular to the fault. Gravity force is presumed to have worked in compression in the vertical direction, so the axis of intermediate principal normal stresses, σ_2, is aligned in this direction.

   In such a model material integrity will not be affected by stresses until the condition of Coulomb's law is met. A rupture will then occur and two families of slip surfaces, that is two sets of fractures, originate. The line of surface intersection coincides with the direction of intermediate principal normal stress. In the model it coincides with the vertical along which the axis of σ_2 is arranged. In the plane of σ_1, σ_2, which coincides with the Earth's surface in our model, the slip lines are in fact rays symmetrical about the σ_1 and σ_2 axes. The axes are bisectrixes: the axis of σ_1 (maximum stresses) which coincides in our model with the fault trend in a bisectrix of the acute angle equal to

$$\pi/2 - \varphi$$

and the axis of σ_3 is a bisectrix of the angle equal to

$$180° - (\pi/2 - \varphi)$$

or in this case 68° and 112° respectively.

These conclusions drawn from the model compared with the actual fracture pattern show that there is no great difference. Indeed, (i) there are two main fracture rays on the Earth's surface; (ii) the fracture planes intersect along the vertical line to within 5°; (iii) the fault trend and direction perpendicular to the fault coincide with bisectrixes to within 5°; (iv) the fault strike brought in line with the axis of maximum principal normal stress σ_1 coincides with the acute angle between the fracture sets, whereas the direction perpendicular to the fault coincides with the obtuse angle; (v) the angle between the sets coincides with the value determined theoretically with $\varphi = 22°$ to within 5°.

Hence the proposed model of loam properties and dynamics of fracturing at the seismic fault agrees well with the observations. It accounts for the two sets of feather joints. Yet the model fails to account for the main fault, fractures parallel to it and feather joints arranged at 68° to the fault. These fractures originated with a somewhat different position of axes of principal normal stresses. Using the Mohr criterion, the axis of maximum principal normal stresses was directed north-eastwards, azimuth being about 50°, at a time when the main fault was being formed. The axis of minimum stress, σ_3, was directed south-eastwards, azimuth 140°, and the axis of intermediate principal normal stress σ_2 was almost perpendicular to the Earth's surface.

Taking into account that the eastern flank of the main Mogod fault is somewhat uplifted and the centres of the feather joint sets (Figure 9.12) are located some distance from the great circle of the diagram, it may be inferred that the axis of σ_2 is somewhat inclined, plunging steeply south-westwards. The other axes of principal normal stresses also deviate from the horizontal position accordingly. Based on seismic data, the positions of the axes in the hypocentre of the Mogod earthquake is similar to the one determined for the epicentre (see Table 9.5).

Thus, the two consecutive states of the rock stress field in the epicentre correspond to the two pairs of conjugate feather shear fractures. The first (in time) stress field is similar to stresses in the hypocentre. Associated with it genetically are fractures parallel to the fault and a conjugate set which forms an angle of 68° with the fault. Once the fault and two sets of Riedel shears are formed, the stresses are removed, giving way to new ones. The axis of maximum principal normal stresses at the second stage of the fault generation is directed along the fault and coincides with the fault plane velocity vector. Its position is apparently determined by friction on the fault planes. The second (in time) stress field gives rise to a pair of feather shear fractures. This pattern was very obvious in the Mogod and Yanitse earthquakes of intensity 10.

The multivariate statistical relationships between earthquake and seismic dislocation parameters are fairly close and, therefore, offer some advantages over their bivariate counterparts. They can be used in processing palaeoseismogeological data with a view to reconstructing the magnitude and intensity of previous earthquakes.

Among the seismic dislocation parameters (fault length, aperture, amount of relative limb displacement) under study, the amount of displacement is most closely related to earthquake intensity and magnitude. It can be treated as a major indicator of the epicentral zone position.

The feather joints in seismic dislocations follow a particular pattern, and using the Mohr criterion of failure, the location of axes of principal normal stresses in the epicentre during earthquakes can be assessed.

As a final comment, Pollard and Aydin (1988, p.1186) consider that

Although a Mohr diagram is useful in representing a homogeneous stress or strain field and providing an empirical failure criterion, it does not represent the heterogeneous fields associated with a joint. As a tool for studying joints, therefore, a Mohr diagram cannot further our understanding of the process of joint initiation, propagation, and arrest. The methods that explicitly treat the heterogeneous fields of fracture, as pioneered by Inglis, Griffith (1921, 1924), and Irwin (1958), should replace this reliance on the Mohr diagram.

Table 9.5 Position of axes of principal normal stresses in Mogod earthquake

Axis	In hypocentre based on seismological data (Bayaraa, 1970)		In epicentre based on geological data	
	azimuth (degree)	inclination (degree)	azimuth (degree)	inclination (degree)
δ_1	62	7	49	0
δ_2	347	42	319	80
δ_3	167	48	139	10

Chapter 10
Hydraulics of water movement in rock masses

10.1   Seepage heterogeneity and seepage models

Groundwater, other liquids and gases enclosed in the Earth's crust flow in pores, fractures and large cavities of suffosive, karstic or some other origin. These three types of channel differ greatly from each other in size, form and distribution pattern. Hydraulic resistances in those channels are, therefore, also different. For instance, in wide channels, say karstic cavities or abandoned mine workings, the seepage flows may be characterized by turbulence, which is not typical of rock channels.

The distance between pores is millimetres or fractions of a millimetre, while that between fractures is tens of centimetres and metres. In karstic and heterogeneous fractured rock masses the distance between seepage channels is tens and hundreds of metres. This fact calls for different models to be constructed for porous, fractured and karstic rock masses.

Porous sands and sandy clays are usually free from fractures. Seepage occurs along small pores which can be spaced a billion times closer than the size of engineering structures. Such ground can therefore be modelled as a continuous permeable medium. The continuous medium model adopted in the classic permeability theory has been effectively used for more than 100 years. As defined by mathematical analysis, the model medium is homogeneous and its parameters are constant. This condition may not agree with geological reality when studying seepage in small volumes of ground similar in size to individual seepage channels.

Statistical analysis shows (Rats and Chernyshev, 1965) that a body can be considered to be large and homogeneous if its linear dimensions are 10^3 times greater than the distance between seepage channels. The continuous medium model can be applied conventionally to a body whose linear dimensions are not too large and exceed the distance between seepage channels by 10–1000 times only. This body should be treated as a homogeneous part of a heterogeneous rock mass. Hence, sand can also be treated approximately as a continuous medium in an instrument for testing from 1 dm^3 of ground. Based on a number of seepage experiments, the arithmetic mean can be calculated of the permeability coefficient, which is an unbiased estimate of mathematical expectation. Not so with testing the same sand in smaller samples.

For the body whose linear dimensions exceed the distance between seepage channels by 3–9 times only, the effect of individual channels becomes great. The arithmetic means of a number of test results leads to biased estimates. Hence, it is impossible to apply the continuous medium model to such bodies. The foregoing shows that the continuous medium model for such ground as sand and clay can be used in calculating permeability at most structures. It is different with rocks and hard soils. With a dense fracture system the seepage channels are spaced 0.1 m apart, and generally their spacing does not exceed 1 m. So the continuous medium model can be applied only to lithologically homogeneous fractured rock masses measuring more than 100 m across. A fractured rock mass measuring 10–100 m across can be roughly treated as a continuous medium. A fractured rock mass with a size of less than 10 m across cannot be taken as a continuous medium. Even if seepage takes place mostly along large tectonic fractures or karstic zones, the continuous medium model can be applied conventionally only to a rock mass measuring more than 1000 m across. Therefore, in cases where fractured rock masses are involved, the continuous medium model often proves ineffective, and models constructed from discrete water-conducting elements are better suited for this purpose.

In reviews of the permeability of fractured rock, the problem may be simplified by considering only parallel arrays of cracks with different openings, and the permeability of parallel arrays of smooth cracks (Hoek and Bray, 1977; Barton, 1987). The permeability of a rock mass is very sensitive to the opening of fractures, and since opening changes with stress, the permeability of the rock mass will be sensitive to stress. Permeability will be largely determined by fractures with the widest opening, as becomes apparent later.

10.2   Fundamentals of the fracture system permeability theory[*]

The tensor permeability theory and the method for determining the permeability coefficient for fractured rock masses has been dealt with in papers by Romm (1966), Snow (1968) and Wittke (1984). The method was developed for continuous fracture systems consisting of an unspecified number of fracture sets. Orientation, density and width of fractures are taken to be constant in each set. This condition, which is in poor agreement with the actual heterogeneity of fracture systems, restricts the application of the tensor theory to practical calculations.

Water permeability is characterized by the symmetric tensor of the second rank k for open fractures (Romm, 1966):

$$
k = \frac{1}{12}
\begin{Vmatrix}
\sum_{i=1}^{n} \dfrac{b_i^3}{a_i}(1-\alpha_{1i}^2) & -\sum_{i=1}^{n} \dfrac{b_i^3}{a_i}\alpha_{1i}\alpha_{2i} & -\sum_{i=1}^{n} \dfrac{b_i^3}{a_i}\alpha_{1i}\alpha_{3i} \\
-\sum_{i=1}^{n} \dfrac{b_i^3}{a_i}\alpha_{2i}\alpha_{1i} & \sum_{i=1}^{n} \dfrac{b_i^3}{a_i}(1-\alpha_{2i}^2) & -\sum_{i=1}^{n} \dfrac{b_i^3}{a_i}\alpha_{2i}\alpha_{3i} \\
-\sum_{i=1}^{n} \dfrac{b_i^3}{a_i}\alpha_{3i}\alpha_{1i} & -\sum_{i=1}^{n} \dfrac{b_i^3}{a_i}\alpha_{3i}\alpha_{2i} & \sum_{i=1}^{n} \dfrac{b_i^3}{a_i}(1-\alpha_{3i}^2)
\end{Vmatrix}
\tag{10.1}
$$

[*] Written in co-authorship with V. V. Semyonov.

and (Chernyshev, 1971):

$$k = k_{fr} \left\| \begin{matrix} \sum_{i=1}^{n}\dfrac{b_i}{a_i}(1-\alpha^2_{1i}) - \sum_{i=1}^{n}\dfrac{b_i}{a_i}\alpha_{1i}\alpha_{2i} - \sum_{i=1}^{n}\dfrac{b_i}{a_i}\alpha_{1i}\alpha_{3i} \\\\ -\sum_{i=1}^{n}\dfrac{b_i}{a_i}\alpha_{2i}\alpha_{1i} \quad \sum_{i=1}^{n}\dfrac{b_i}{a_i}(1-\alpha^2_{2i}) - \sum_{i=1}^{n}\dfrac{b_i}{a_i}\alpha_{2i}\alpha_{3i} \\\\ -\sum_{i=1}^{n}\dfrac{b_i}{a_i}\alpha_{3i}\alpha_{1i} - \sum_{i=1}^{n}\dfrac{b_i}{a_i}\alpha_{3i}\alpha_{2i} \quad \sum_{i=1}^{n}\dfrac{b_i}{a_i}(1-\alpha^2_{3i}) \end{matrix} \right\| \qquad (10.2)$$

for fractures filled with loose material, where

k_{fr} is the permeability coefficient of fracture filler, cm/s;
n is the number of fractures in the rock mass;
i is the ordinal number of the fracture set;
b is average fracture width in the set, cm;
a_i is fracture spacing in the set, cm;
α_{1i}, α_{2i}, α_{3i} are direction cosines of vector perpendicular to the ith fracture set in the cordinate system x, y, z, (x – east, y – north, z – zenith).

For any direction coinciding with that of the flow gradient the permeability coefficient of a fractured rock mass can be calculated with the following formula:

$$k = \frac{g}{12v}\sum_{i=1}^{n}\frac{b_i^3}{a_i+b_i}\{1-[\sin\beta_i(\cos\alpha_i\cos\varphi+\sin\alpha_i\cos\varphi)+\cos\beta_i\cos\gamma]^2\}$$

$$+ \sum_{i=1}^{n}\frac{b_i k_{fr}}{a_i+b_i}\{1-[\sin\beta_i(\cos\alpha_i\cos\varphi + \sin\alpha_i\cos\varphi)+\cos\beta_i\cos\gamma]^2\}\,(10.3)$$

The formula can be applied to a rock mass with open and filled fracture sets (Figure 10.1), where

k is permeability coefficient of rock mass, cm/s;
g is gravitational acceleration, cm/s^2;
v is kinematic viscosity factor, cm^2/s;
α_i is dip azimuth of the ith fracture set;
β_i is dip angle of the ith fracture set;
ρ, φ, γ are angles between the direction for which permeability coefficient is assessed and coordinate axes x, y and z;
k_{fr} is permeability coefficient of fracture filler, cm/s.

If a rock mass can be assumed to consist of cubes with edge a separated by means of fractures with width b,

$$k = \frac{gb^3}{\sigma v a} \qquad (10.4)$$

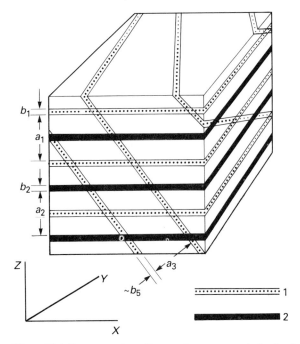

Figure 10.1 Fracture system diagram for tensor analysis: 1 – fracture-bed; 2 – fracture-fissure; a_1, a_2, a_3 – fracture spacing in sets 1, 2 and 3; b_1, b_2, b_3 – fracture width in the same sets

holds true in the case where there is no filler in fractures of any trend (Romm, 1966).

With a loose fracture filler with permeability coefficient k_{fr} the same rock mass has

$$k = \frac{2k_{fr}\cdot b}{a+b} \tag{10.5}$$

All the above expressions were written on the assumption that the permeability coefficient of a fracture filler is the same for all fracture sets. If this is not true and each fracture set is characterized by its particular permeability coefficient of a fracture filler $(k_{fr})_i$, the latter is under the summation sign. For instance, for the vertical direction:

$$k = \frac{g}{12v} \sum_{i=1}^{n} \frac{b_i^3}{a_i+b_i} \sin^2\beta_i + \sum_{i=1}^{n} \frac{(k_{fri})b_i}{a_i+b_i} \sin^2\beta_i \tag{10.6}$$

The above expressions (10.3)–(10.6) enable the permeability of a rock mass to be determined. To obtain the rock mass permeability coefficient, the permeability coefficient characterizing the porous permeability in a rock sample has to be combined with the permeability coefficient characterizing fractured permeability, which was experimentally tested on a model of a fractured-porous medium (Wilson and Witherspoon, 1974).

The method of linear elements is intended for calculating water flow rate, velocity and head pressure in fractured rock masses (Wittke and Louis, 1966; Louis, 1968; Wittke, 1970, 1984). It can be applied to fracture systems of different configuration except non-persistent systems. Calculations are performed on the rock mass model shown in Figure 10.2. A plane section of a rock mass is considered. Fractures in the section are shown as intersecting lines. Each linear element of a fracture system has its individual features. Length and average width are to be determined for such an element. In the event of a loose fracture filler, its permeability coefficient is to be determined. If the fracture wall roughness is great, the roughness parameters are to be found. The problem consists of determining the flow rate of a particular rock mass at a predetermined head pressure and of calculating velocities and piezometric surfaces in the elements of the fracture system.

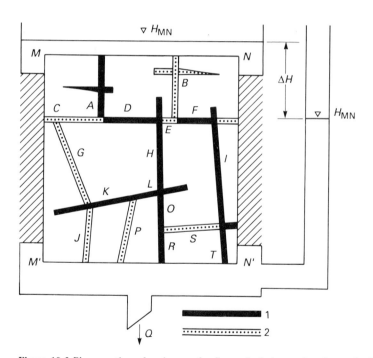

Figure 10.2 Plane section of rock mass for flow calculations using the method of linear elements

The solution is as follows. Denote the specified level drop between the permeable boundaries MN and $M'N'$ of the rock mass by ΔH. The rock mass boundaries MM' and MN' are assumed to be pervious. Now denote the total flow rate from the boundary MN to the boundary $M'N'$ per unit time by Q. The hydraulic slope in the rock mass in question will be

$$I = \frac{\Delta H}{MN'} \tag{10.7}$$

The overall flow rate of the rock mass is the total of flow rates of elementary fractures in a section perpendicular to the pressure gradient. For instance

$$Q = q_c + q_b \tag{10.8}$$

where q_c and q_b are flow rates of linear elements C and B.

It is obvious that the flow rate of some particular element can be determined only in combination with those of all other elements. All the water flow parameters are closely interrelated in the rock mass model as they are in the actual rock mass. So it is necessary to form a set of equations to describe water flow in all the elements at one time. The system comprises three groups of equations. Equations of the first group reflect the condition of flow continuity in each fracture set node. From this it is inferred that in each fracture set node the amount of inflowing and outflowing water forms the algebraic sum of zero. The flow entering the node is taken as positive, while the outgoing flow is taken as negative. This condition can be written for all the nodes and hence we have a set of n equations:

$$\sum_{N_p} q_i = 0 \qquad (\text{for } p = 1, \ldots, n) \tag{10.9}$$

where N_p is a totality of linear elements of the fracture system intersecting at a given node with the number p.

The second group of equations is derived on the assumption that the piezometric surface is continuous for a given path. This implies that the algebraic sum of the head pressure drop is also equal to zero:

$$\sum_{R_j} l_i I_i = 0 \qquad (\text{for } j = 1, \ldots, m) \tag{10.10}$$

where

R_j is a totality of linear elements outlining the polygon with the number j;
l_i is length of linear element;
I_i is flow gradient within linear element;
I is generalized flow gradient in rock mass in the portion R_j.

The polygons are not closed along the boundaries MN and $M'N'$ with specified boundary conditions. For instance, a polygon of the elements J, K and P is not closed in Figure 10.2. For such polygons we can write, instead of (10.10),

$$\sum_{R_h} l_i I_i = \varphi_h - \varphi_{h+1} \qquad (\text{for } h = 1, \ldots, r) \tag{10.11}$$

where

R_h is a totality of linear elements in polygon located on rock mass contour;

h is polygon number;

$\varphi_h - \varphi_{h+1}$ is the difference in pressure surfaces between fracture terminations in rock mass contour (e.g. the difference in pressure surfaces between the terminations of fractures J and P in the boundary $M'N'$ in Figure 10.2

Finally, the boundary conditions specified on the rock mass contour should be interrelated by the equation. For example, in the range from the beginning of the linear element B to the end of the element T (see Figure 10.2) the total of surface drops between the fracture set nodes is the overall drop in surfaces between the model boundaries

$$l_B I_B + l_F I_F + I_I + l_T I_T = H_{MN} - H_{M'N'} \tag{10.12}$$

where

H_{MN} is water pressure surface at the boundary MN;
$H_{M'N'}$ is water pressure surface at the boundary $M'N'$

The expressions (10.9)–(10.12) form a set of $n + m + r + j$ equations. In this set the unknowns are only flow gradients I_i for each linear element of the fracture system. The flow rates q_i can be expressed in terms of the known geometric parameters of the fractures. So we have a set of $n+m+r+j$ linear equations. The number of equations is equal to that of the unknowns I_i. Solving the set of equations will determine the hydraulic slope I_{io} for each fracture, and so verify the hypothesis on laminar water flow for all fractures. If laminar flow takes place in all unfilled fractures, the solution I_{io} holds true for all open and filled fractures. Now from equation (10.8) the total flow rate of the rock mass can be calculated and then the rock mass permeability coefficient can be determined from Darcy's law on the basis of such parameters as flow rate, surface drop and section of the rock mass as a continuous medium. In designations (see Figure 10.2) Darcy's law shall be written for the plane-horizontal flow section as

$$k = \frac{QMM'}{\Delta HMN} \tag{10.13}$$

where

k is the permeability coefficient of the fractured rock mass;
Q is total flow rate through the section and any parallel section;
ΔH is the difference in surfaces between the sections MN and $M'N'$.

If the I_{io} solution shows that turbulent flow may take place in some fractures, those fractures should be characterized by the non-linear relationship between flow gradient and flow rate. The new set comprising non-linear equations is only solved by the iteration method.

The solution of the set of equations makes it possible not only to determine the rock mass permeability coefficient; flow velocities also obtained from the equations are of no less importance. Knowing these velocities, suffosive resistance of some portions of the rock mass can be assessed to outline potential washout areas.

Sometimes, natural water-conducting objects are similar in form to a tube rather than to a fracture. For linear elements of a tubular form, use is made of hydraulics equations to calculate flow rate in a rough tube (Ivannikova, 1988).

The method of finite elements is an effective numerical method for solving a variety of engineering and physical problems, among others, in structural mechanics, hydromechanics, mechanics of continua. In some papers (Dershowitz and Einstein, 1987, Elsworth and Piggot, 1987), the method of finite elements is used in computational modelling of water movement in fracture systems. The method deals with water flow through a rock mass composed of porous pervious fractured rocks. The method is based on the discretization of a particular area and its representation as a set of finite elements: spatial, plane or one-dimensional. The properties of the medium within each particular element can be specified individually, which makes it possible to allow for seepage heterogeneity and anisotropy of an intrablock space, as well as for the different permeabilities of fracture portions.

The problem can be simplified by making it a plane one and considering water flow in a continuous fracture system, assuming the intrablock space to be impervious. The diagram of the fractured rock mass is shown in Figure 10.2. The fracture system is represented as a set of N_{el} segments – one-dimensional finite elements connected at nodal points whose number is denoted by N. Each element is characterized by length l, average width b and, provided there is a loose filler, by its permeability coefficient k_{fr}. The above parameters may vary for individual elements

The method is aimed at establishing major relationships for individual elements, in this case relations between nodal head pressures and flow rates.

Consider a binodal symplex-element with the nodes i and j arranging it in the coordinate system OX in line with the element axis. The distribution of the head pressures $h_{(e)}$ along the length of the element are linear with respect to the coordinate x:

$$h_{(e)} = \alpha_1 + \alpha_2 x \tag{10.14}$$

According to (10.14), the head pressures H_i and H_j in the element nodes are determined by the relations

$$\begin{aligned} H_i &= \alpha_1 + \alpha_2 X_i \\ H_j &= \alpha_1 + \alpha_2 X_j \end{aligned} \tag{10.15}$$

wherefrom the coefficients α_1 and α_2 can be obtained:

$$\alpha_1 = \frac{H_i X_j - H_j X_i}{l}$$

$$\alpha_2 = \frac{H_j - H_i}{l} \tag{10.16}$$

In this case the expression (10.14) is reduced to the following form:

$$h_{(e)} = \frac{H_i X_j - H_j X_i}{l} + \frac{H_j - H_i}{l} x \qquad (10.17)$$

or, using form functions N_i and N_j,

$$h_{(e)} = N_i H_i + N_j H_j \qquad (10.18)$$

where

$$N_i = \frac{X_j - x}{l}$$

$$N_j = \frac{x - X_i}{l} \qquad (10.19)$$

In matrix form the relation (10.18) can be represented as

$$h_{(e)} = [N]\{H\} \qquad (10.20)$$

where

$[N] = [N_i N_j]$ is a row matrix of form functions;

$\{H\} = \begin{Bmatrix} H_i \\ H_j \end{Bmatrix}$ is a column matrix of nodal head pressure values.

In the event of the plane problem the flow rate q through an element of the fracture filled with some porous material with laminar seepage is determined by the relation

$$q = -k_{fr} b J_{(e)} \qquad (10.21)$$

where $J_{(e)}$ is hydraulic gradient

$$J_{(e)} \frac{dh_{(e)}}{dx} \qquad (10.22)$$

or writing the expression (10.22) and substituting (10.18):

$$J_{(e)} = \frac{dN_i}{dx} H_i + \frac{dN_j}{dx} H_j = \frac{1}{l}(H_j - H_i) = \frac{1}{l}[-1 \quad 1]\{H\} \qquad (10.23)$$

Here it is possible to obtain the relations in a simple way. Since with a given function of head pressures the hydraulic gradient $J_{(e)}$ and, hence, the flow rate q are constant along the length of the element, the flow rates q_i and q_j in the element nodes may be calculated assuming the flow outgoing from the node i into the element ($H_i > H_j$) to be positive and the one entering the node j to be negative:

$$\{q_{(e)}\} = \frac{k_{fr} \cdot b}{l} \begin{pmatrix} 1 & -1 \\ -1 & 1 \end{pmatrix} \{H\} \qquad (10.24)$$

where $\{q_{(e)}\} = \begin{Bmatrix} q_i \\ q_j \end{Bmatrix}$ is a column matrix of nodal flow rates.

In the traditional form the expression (10.24) is as follows

$$\{q_{(e)}\} = [p_{(e)}]\{H\} \tag{10.25}$$

where

$$[p_{(e)}] = \frac{k_{fr} \cdot b}{l} \begin{bmatrix} 1 & -1 \\ -1 & 1 \end{bmatrix} \tag{10.26}$$

The square symmetric matrix $[P_{(e)}]$ interrelating the nodal head pressures and flow rates in the element is referred to as the element permeability matrix.

It may be shown that with the laminar steady-state flow through a smooth open fracture the expression for the element permeability matrix is represented as

$$[p_{(e)}] = \frac{g \cdot b^3}{12 \cdot l \cdot v} \begin{bmatrix} 1 & -1 \\ -1 & 1 \end{bmatrix} \tag{10.27}$$

The equation (10.25) written for an individual element can be applied to a set of elements connected at N nodes. In this case it is reduced to a similar form:

$$\{Q\} = [P]\{H\} \tag{10.28}$$

where

$\{Q\}$ is an N flow rate column matrix;
$[P]$ is an $N \times N$ permeability matrix;
$\{H\}$ is an N head pressure column matrix.

The permeability matrix $[P]$ is a positive definite symmetric matrix with a banded diagonal structure. Its terms are determined by the summation of those of the permeability matrices of individual elements $[p_{(e)}]$.

In expanded form the matrix expression (10.28) is a set of N linear equations which can be solved by standard methods under specified conditions.

The set (10.28) is characterized in that either head pressure or flow rate is the unknown for each node. In the problems under consideration the flow rates in nodes located on impervious boundaries are equal to zero (the algebraic sum of the outgoing and incoming flows is equal to zero). So it is nodal head pressures that are the unknowns to be determined. In the contour nodes corresponding to the inlet and outlet of the seepage flow, it is nodal head pressures that are generally specified, flow rates being the unknowns. In either case the set comprises N linear equations in N unknowns.

The solution of the set of linear equations determines nodal head pressures and flow rates based on which hydraulic gradients and seepage velocities can be calculated for each particular element.

The rock mass permeability coefficient is determined with the following formula:

$$k = \left(\sum_{i=1}^{n} Q_i \right) \Big/ SJ \tag{10.29}$$

where

> n is the number of nodes on inlet contour;
> Q_i is flow rate in the ith node on inlet contour;
> S is rock mass cross-section;
> J is hydraulic flow gradient in the rock mass area under study.

The hypothesis on laminar flow which is generally taken as a first approximation should be checked for each finite element following the calculations. This operation can be performed, say, by determining Reynolds numbers Re in the elements and correlating them with critical Reynolds numbers Re_{cr}. In the event of $Re < Re_{cr}$ for a totality of elements, the linear flow laws hold true for each particular element and, hence, for the entire set. So the solution obtained is final. Otherwise, with at least some of the Reynolds numbers exceeding critical values, the non-linear relationships between hydraulic gradient and flow rate (see Section 10.6) are to be taken for appropriate elements. In such an event, the set of equations (10.28) becomes non-linear to be solved by the iteration method.

The advantage of the method of finite elements over that of linear elements resides in the fact that the former eliminates the need for preliminary analysis of fracture set hydraulics, thus making three different groups of equations redundant. The method of finite elements calls for a set of equations to be formed by the standard technique on the basis of the formalized procedure.

10.3   Studies of water flow in an isolated fracture

Prior to studying water flow in a set of intersecting and interacting fractures, insight has to be gained into water flow in an isolated fracture. This process has been little studied in actual rocks. Investigations have been performed but mostly under simulated conditions (Lomize, 1951; Huitt, 1956; Romm, 1966; Louis, 1968; Zhilenkov, 1975). The results of those investigations can be summarized as follows. The model of an actual rock fracture is based on a slot fissure with smooth parallel walls, the distance between which can be taken as the one equalling the average distance between the rough walls of the fracture. A fracture is irregular in form, which increases its hydraulic resistance. This fact can be allowed for through corrections applied for wall roughness. All these points have been ascertained, it is thought, with a high reliability and need no verification, except for wall roughness. The relationship between hydraulic resistance and fissure wall roughness is viewed from different angles by many authors. Yet the calculations have to be based on some particular empirical relation.

In models of a rough fracture proposed by Lomize and Louis, the hydraulic resistance depends on the absolute height of an elevated portion on the rough surface of the fracture wall and on fracture curvature. In actual fractures all the elevated portions are of different height and width and hence it is practically impossible to distinguish between curvature and roughness. To use the results of the experiments conducted by Lomize and

Louis, roughness was determined as follows. The difference in fracture opening at neighbouring measuring points was taken as a height of the elevated portion. The measurements were taken at a spacing of 1 mm. The absolute roughness was calculated using the formula

$$e' = \frac{1}{n} \sum_{i=1}^{n} |b_i - b_{i+1}| \tag{10.30}$$

where

  e' is absolute roughness;
  n is number of measurements;
  b_i is opening at the ith measuring point.

Based on such an assumption, one of the fracture walls is taken as smooth and the other as rough. Assuming the walls bend towards each other, the absolute roughness is determined as

$$e' = \frac{1}{n} \sum_{i=1}^{n} \frac{|b_i - b_{i+1}|}{2} \tag{10.31}$$

The relative roughness is determined as a ratio of the average height e of the elevated portion to average fracture opening. The absolute and relative wall roughnesses were determined on the basis of measurements and used below when comparing models of an isolated fracture.

There are several fracture and water flow models, and the most important of these are:

1. Relationship for a smooth fissure:

$$q = \frac{gb^3}{12v} I \tag{10.32}$$

where

  q is water flow rate in fracture, cm^3/s;
  b is average fracture width, cm;
  g is gravitational acceleration, cm/s^2
  v is water kinematic viscosity coefficient, cm^2/s;
  I is hydraulic gradient of flow in fracture.

2. Relationship for a rough fissure (Lomize, 1951):

$$q = \frac{g}{12v} b^3 I \frac{1}{\left[1 + 6\left(\dfrac{e}{b}\right)^{1.5} \right]} \tag{10.33}$$

The relationship is recommended for fissures with $e/b > 0.066$. In the present case this condition is met.

3. Relationship for a rough fissure (Louis 1968):

$$q = \frac{g}{12\nu}b^3 I \cfrac{1}{1 + 8.8\left(\cfrac{e}{2b}\right)^{1.5}} \tag{10.34}$$

Those relations (10.32)–(10.34) were compared with experimental data (Chernyshev, 1979). For comparison, flow rate–head pressure relationships and such parameters as width and roughness were determined for three fractures. The experimental data are correlated with calculated ones (Figure 10.3 and Table 10.1). In calculations the relative roughness was taken both as for formula (10.30) and formula (10.31).

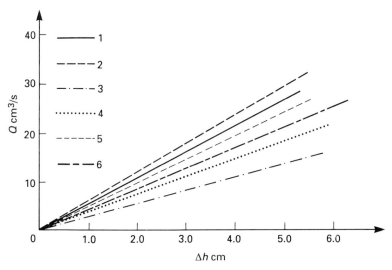

Figure 10.3 Relationship between flow rate and difference in piezometric heads at fracture ends for specimen 2: 1 – based on experiment results; 2 – calculated from the formula for a smooth fissure; 3 – calculated from Lomize's formula with $e' = b_{i+1} - b_i$; 4 – calculated from Louis' formula with $e' = b_{i+1} - b_i$; 5 – calculated from Louis' formula with $e = e'/2$; 6 – calculated from Lomize's formula with $e = e'/2$

Table 10.1 Relative errors of calculated data for different calculation methods %

		Relations			
Specimens	Smooth fissure	Lomize e'	Lomize e	Louis e'	Louis e
1	14.2	21.4	1.2	7.5	5.3
2	9.9	48.9	21.6	30.8	9.2
3	19.2	68.6	48.2	55.7	37.2
Average	14.4	46.3	23.7	31.3	17.2

As regards the agreement between the experimental and calculated data for specimen No. 1, the best results were obtained with the help of the Lomize and Louis relationships with the relative roughness determined from (10.31). These relationships also proved most effective for experiments with specimen Nos. 2 and 3 (see Table 10.1).

In conclusion it should be noted that with roughness allowed for, there was not a higher accuracy of forecasting water flow in a fracture with preset geometrical parameters. The formula taken for a smooth fissure gives a flow rate on the high side, but the error is not large. So in subsequent calculations use was made of the formula for a smooth fissure.

10.4 Laboratory and field studies of water movement in rock fracture systems

Laboratory experiments

Studies of water flow in fracture systems are made to check the methods for calculating water permeability based on the geometrical parameters of fractures. At the same time the measurements also covered parameters of the flow running through a fractured rock mass and geometrical parameters of water-conducting fractures. The experiments were run under laboratory and field conditions.

The laboratory experiments involved physical modelling of the groundwater flow through a fractured rock mass with a similarity factor for all the parameters of the process under study equalling one. The pressure water flow in marble and granite was studied for three samples assigned Nos. 4, 5, 6.

Rock sample description

Sample no. 4 was made of a quadrangular marble plate measuring 40×55 cm with an average thickness of 7.8 cm. The plate was broken into 14 blocks which differed in size (Figure 10.4). Tension cracks resulting from the plate dissection featured fairly rough wall surfaces and were curved in form, thus simulating well the form of a natural tension fracture. No attempt was made to generate smooth cracks similar to natural shear fractures since the water flow theory is obviously more applicable to such fractures than to complex-shaped tension fractures.

The fracture system model, much like a natural fracture system, was made with no loss of material resulting from breaking the plate. Under natural conditions the formation of tension fractures in a rock mass does not apparently involve the removal of fracture material, and hence the two walls of natural fractures represent a sort of die and an imprint. Opening results from some minor longitudinal displacement of the fracture walls and their subsequent closing with contact at individual points. In the model the total volume of marble grains amounted to some 3 cm³ or about 0.01% of the plate volume and not more than 5% of the fracture volume. Thus, full equality of the fracture wall surfaces was achieved in modelling. The form of fractures with such surfaces was similar to that of natural fractures.

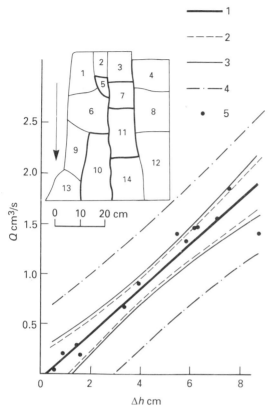

Figure 10.4 Diagram of water-conducting fractures and $Q(\Delta h)$ graph for sample 4c. Water-conducting fractures are shown as a bold line, and blind fractures as a thin line. For the graph: 1 – regression line; 2 – confidence regression zone; 3 – tolerance limits; 4 – confidence intervals for individual values; 5 – experimental points

Arrangement and intersection of fractures in a sample also reflected the natural environment. As is usually the case under natural conditions, fractures in sample no. 4 grouped into sets almost parallel to each other. In one of the sets, shown as the vertical one, fractures are almost continuous (two out of three intersecting the sample). The two fracture sets correspond to two consecutive generations. In the first they are longer and more persistent in direction, whereas in the second generation they are shorter, less parallel and spaced more widely. These features of artificial fractures agree with those of natural fracture systems. When making the sample, an attempt was made to reproduce natural conditions of fracturing rather than to develop a laboratory model reflecting properties of the mathematical model of the object. It would be ideal to run experiments in a rock mass. However, it is only under laboratory conditions that the difficulties relating to studies of geometry of all the fractures in the system could be overcome.

The sample was positioned in a flume for studying plane pressure flow. The blocks were glued to the bottom of the flume. Openings were measured individually for each fracture at 6–8 points. Use was made of a feeler gauge with a set of plates ensuring an accuracy of at least 0.01 cm, but even so wall surface roughness and curvature made it impossible to achieve such a high accuracy. Each particular measurement exhibited an error of about 0.02 cm. The standard error in assessing average fracture opening was taken as 0.01 cm. The average width of some fractures reached 1.5–2.0 mm. In natural rock masses, except in the hypergene zone, fractures are usually much narrower. So the test sample simulated a zone of weathered rocks featuring low density.

The length of fractures was measured with an error not exceeding 0.2 cm.

Thus all the sample parameters were obtained to calculate permeability coefficient based on fracture parameters (Table 10.2). At the same time the sample was transformed into the model of a bed interposed between two confining layers and suitable for studying pressure groundwater flow.

Sample no. 5 was also made of marble. The plate measured 42×70 cm and was about 10 cm thick. The form of the fracture set is shown in Figure 10.5.

Table 10.2 Parameters

	Designations	4a	4b	4c	4e
Arithmetic mean	Δh	2.525	4.985	4.154	5.300
Arithmetic mean	$\bar{Q}$	4.559	0.398	0.873	1.982
Standard deviation Δh	$S_{\Delta h}$	1.747	2.011	2.631	2.724
Standard deviation Q	S_Q	3.106	0.301	0.617	1.013
Free term of regression equation	$\hat{a}$	0.096	−0.326	−0.057	0.077
Angular coefficient of regression equation	$\hat{b}$	1.767	0.145	0.224	0.359
Correlation coefficient	r	0.994	0.972	0.955	0.966
Standard forecast error	$S_{Q\Delta h}$	0.350	0.076	0.192	0.272
Standard errors of regression equation coefficients	{ $S_{\hat{a}}$	0.101	0.027	0.053	0.085
	{ $S_{\hat{b}}$	0.060	0.014	0.021	0.023
Confidence boundaries of correlation coefficient at 0.05 reliability level	{ r_1	0.998	0.994	0.986	0.989
	{ r_2	0.977	0.828	0.840	0.880
Confidence intervals for regression equation coefficients	{ $a \pm t_{\alpha/2} \cdot S_{\hat{a}}$	−0.129	−0.392	−0.174	−0.089
		0.322	−0.260	0.060	0.243
	{ $\hat{b} \pm t_{\alpha/2} \cdot S_{\hat{b}}$	1.632	0.110	0.178	0.296
		1.902	0.180	0.270	0.423

Sample no. 6 was made of coarse-crystalline pink–grey granite. The feldspar crystals in the rock reached 15–20 cm in size and those of hornblende and pyroxene measured 3–5 mm. A quadrangular plate measuring 70×105 cm with a thickness of some 4.0 cm was used. The plate was broken into 34 blocks of different size (Figure 10.6). Due to the high brittleness of the material, it was impossible to avoid some fragments when breaking down the plate. Width, length and orientation of fractures were measured when placing the blocks into the flume. The fracture wall roughness in granite sample no. 6 is similar to that in sample nos. 1–3. The arrangement of fractures in sample no. 6 is much more complex than in the

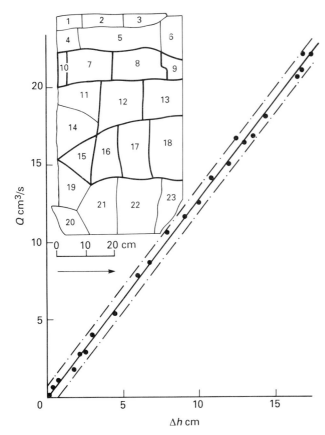

Figure 10.5 Diagram of water-conducting fractures and $Q(\Delta h)$ graph for sample 5. For symbols see Figure 10.4

other samples. There are four fracture sets running along the long and short sides of the sample, as well as along its diagonals. Such an arrangement of fractures is commonly found in granites and is fairly typical of layered rocks in fold limbs.

Procedures

The flume with samples was provided with four tanks to fit the number of sides of the test samples. This made it possible to assess the permeability coefficient of the samples with fractures for different directions. Almost all the fractures on the two parallel lateral surfaces were glued up. Tanks with different water levels were provided at the other two lateral surfaces. The lower tank level was maintained above the upper plane of the plate. This was responsible for the water pressure flow in fractures. Flow stabilization was carried out for 15–30 minutes at high flow rates and within a period of time of up to 4 hours at minimum flow rates. Then a flow rate was measured and a level drop recorded. The levels were changed, the flow

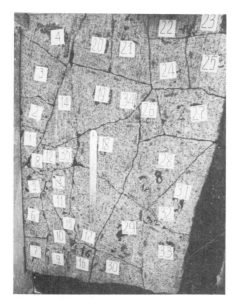

Figure 10.6 Sample 6 in a flume, rule 30 cm long

rate and levels were stabilized again and new measurements taken. Water was recycled into the upper tank through a filter. Prior to the experiment the water was kept in an open vessel at room temperature for at least 24 hours, thus ensuring equilibrium between atmospheric air and water-soluble air. The experiments were run at a temperature of about 20 °C. The fracture set in model no. 4 was changed by sealing the inlet and outlet openings of individual fractures. In modification 4b one fracture was left at the inlet to the sample between blocks 3 and 4. There was also one fracture left between blocks 10 and 14 at the outlet. In modification 4c, fracture 10–13 was additionally opened at the outlet so that the water-conducting set incorporated three more fractures at the contact between block 10 and blocks 6, 9, 13. In subsequent modifications the fracture system was made even more complex (Figure 10.7).

Results

The relationship btween flow rate and level drop is generally linear (Figures 10.4–10.9), which indicates that Darcy's law prevailed in the above experiments. The spread of experimental points on the plots was large in a number of cases. The experimental data were statistically processed according to the regression analysis program to show that the above spread of points does not markedly affect the final result, i.e. determination of the sample permeability coefficient with a specified configuration of the fracture system. For the four modifications of the experiment, confidence regression equation zones were calculated for the average interval values in addition to the linear regression equations. The basic result of the experiment, namely a regression equation, is written as

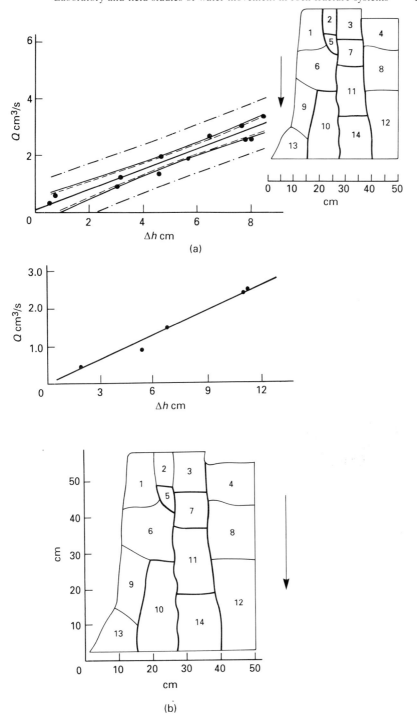

Figure 10.7 Diagram of water-conducting fractures and $Q(\Delta h)$ graph: (a) for sample 4e; (b) for sample 4d. For symbols see Figure 10.4

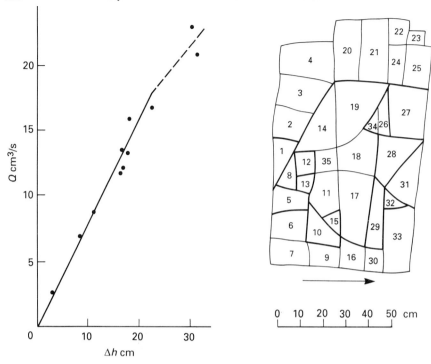

Figure 10.8 Diagram of water-conducting fractures and $Q(\Delta h)$ graph for sample 6b. For symbols see Figure 10.4

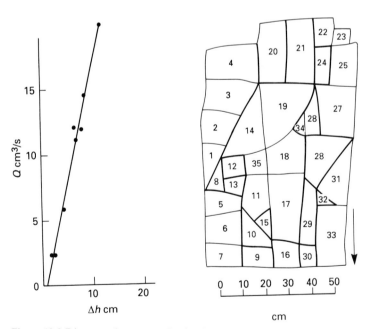

Figure 10.9 Diagram of water-conducting fractures and $Q(\Delta h)$ graph for sample 6c. For symbols see Figure 10.4

$$Q = \hat{a} + \hat{b}\Delta h \tag{10.35}$$

In particular, the result of the experiment 5a is as follows:

$$Q = -0.058 + 1.268\Delta h \qquad r = 0.999 \tag{10.36}$$

The flow rate–gradient relationship is always fairly close. The correlation coefficient at the 5% significance level exceeds 0.8, and the free member of the regression equation ($\hat{a}$) at the same level did not differ from zero, i.e. the $Q\,(\Delta h)$ plot passes through the origin. The correlation coefficient is always assessed at more than 0.95, which points to a close relation between Q and Δh established in the course of the experiments. The proportionality factor $\hat{b}$ in the regression equation characterizes the rock mass permeability coefficient and the error $s_{\hat{b}}$, the error in assessing the permeability coefficient. In accordance with Darcy's law

$$Q = \frac{k \cdot S}{L}\Delta h \tag{10.37}$$

holds true for each model. At the same time, from the experiment it follows that

$$Q = \hat{b}\Delta h \tag{10.38}$$

where

S is area of sample section in the direction perpendicular to water flow direction, cm^2;
L is sample length in the direction of water flow;
k is permeability coefficient, cm/s.

Hence the permeability coefficient of the sample as a continuous medium equals

$$k = \frac{\hat{b} \cdot L}{S} \tag{10.39}$$

The standard error in the experimental determination of the sample permeability coefficient is

$$S_k = \frac{s_{\hat{b}}L}{S} \tag{10.40}$$

Table 10.3 lists calculated values of sample permeability coefficient and errors. The latter are fairly small.

The experiments have shown that the samples feature high permeability coefficients typical of zones of weathering. Sample permeability anisotropy agrees well with that of actual rock masses.

Field experiments

Filling of test holes was undertaken side-by-side with graphic representation of fractures. The experiment was aimed at obtaining data for correlating rock mass permeability coefficients calculated on the basis of fracture parameters with the results of their direct hydrodynamic determi-

Table 10.3 Permeability coefficients of fracture samples

Experiment no.	C_p (cm/s)	S_c (cm/s)	C_p (m/day)	S_c (m/day)
4a	0.19	0.6×10^{-2}	160	5.0
4b	2.2×10^{-2}	0.2×10^{-2}	19	1.6
4c	3.5×10^{-2}	0.3×10^{-2}	30	3.0
4d	3.5×10^{-2}	–	30	–
4e	5.8×10^{-2}	0.3×10^{-2}	50	3.0
5a	0.20	1.2×10^{-2}	172	10.0
5b	5.8×10^{-2}	–	50	–
6a	0.67	–	590	–
6b	0.20	–	172	–
6c	0.33	–	285	–
6d	0.26	–	221	–

nation. The area of field studies is situated in the interfluve of the Orkhon and Selenga in Mongolia. Igneous rocks of different composition (both effusives and intrusives) are overlain by a Quaternary cover 0.5–50 m thick. Active tectonic movements throughout geological history give rise to a dense system of fractures which evolves at present under the influence of earthquakes, tectonic tension and physical weathering.

In the upper part of the rock mass where the experiments were run, fractures fully separate individual blocks. The maximum width of the fractures is 1 cm. The fracture spacing is 10–15 times as large as the fracture opening. There is generally no filler in the fractures. The rock mass in the area under study features small blocks and high porosity. The groundwater surface is clearly defined by holes to be at a depth of 50–70 m. Test pits were sunk manually to a depth of 1–3 m. All visible fractures were represented graphically. A case in point is shown in Figure 10.10. Special emphasis was placed on accurate measurements of fracture width and on description of the composition and density of a loose fracture filler if any. Fractures and portions thereof filled in with hydrothermal minerals were not taken into account.

The filling results (Table 10.4) are compared with results of assessing water permeability on the basis of fracture system geometry. When assessing the difference between the calculated and experimental values, the following factors are taken into consideration: (i) permeability coefficients are determined for the vertical direction; (ii) the accuracy of determining permeability coefficients is not high due to rock mass heterogeneity and small sump area; (iii) a decrease in water permeability with

Table 10.4 Results of filling holes with water

Hole no.	Sump area (m²)	Constant flow rate (m³/day)	Permeability coefficient (m/day)
1501	0.27	58.6	120.0
1502	0.19	5.0	14.0
1503	0.17	2.7	7.0
1507	0.22	2.7	6.0
1509	0.42	0.27	0.4

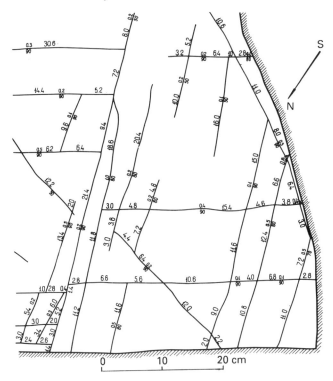

Figure 10.10 Graphic representation of sump (fragmentary view) for filling hole No. 1507 with water: 1 – sump boundary; 2 – fracture line with parameters (numerator-opening, mm and length, cm; denominator – dip angle, degree)

depth could result in a lower value than rocks exposed in the sump. Thus, the null experiment error is far greater than the laboratory experiment error from the preceding section.

Along with filling of test holes, water injections into holes were made to assess the accuracy of the calculation of the permeability coefficient on the basis of fracture parameters. For this purpose fractures were studied in the area of Rogunskaya Hydro on the Vakhsh River. The rock mass in question is composed of Lower Cretaceous red beds. The modern tectonic setting is characterized by horizontal compression of the rock mass. So the fracture opening is a negligible quantity. Large fractures are filled in with mylonite and fault breccia. On the whole the rock mass features fairly low permeability.

Fan core rotary-percussion drilling was employed to drill holes 105 mm in diameter to depths of 80–100 m. The cluster drilling in different directions enabled determination of permeability anisotropy of the rock mass on the basis of results of permeability testing of bore holes. Testing was done by injecting water into individual hole sections. Water permeability of the rock mass in the area adjacent to the opening was determined in greatest detail. Results of injections into the first section, if viewed from the hole mouth, were used. These data are of interest since they apply to the rock mass whose fracturing can be observed in the opening. Fractures in openings near the hole clusters were represented graphically in accordance with the requirements of the method of linear elements. The data thus obtained can also be used in calculating the permeability coefficient with the tensor method.

As a result, for a number of areas of the foundation of the Rogunskaya Hydro, adequate information on fracturing enabled calculation of the permeability coefficient, and the experimental data on water permeability can also be compared with the calculated ones. Should there be any discrepancy between the experimental and calculated data, the following should be taken into account: (i) based on injections, similar to filling test holes, the permeability coefficient error is great; (ii) as opposed to laboratory experiments, field injections fail to bring the calculated and tested fracture set into perfect agreement.

On the whole, the experimental data provided herein enable the permeability coefficient of a fractured rock mass to be calculated by the method of linear elements and by the tensor method. The values thus obtained can be correlated with each other and with results of experimental assessment of the permeability coefficient.

10.5   Accuracy of calculation of the permeability coefficient on the basis of fracture parameters

Random errors in calculating the permeability coefficient by the tensor method

Errors in measuring fracture parameters were dealt with in Section 5.3. Consider now how they affect the accuracy of the calculated permeability coefficient.

The result would have been precise if the arguments in formulae (10.3) and (10.6) had no great fluctuations due to natural heterogeneity of the object and measuring errors.

Actually, fracture parameters (orientation, density/spacing, width) in formulae (10.3), (10.6), are to be treated as random values with known root-mean-square deviations σ or coefficient of variation V. Gravitational acceleration, water kinematic viscosity coefficient, direction cosines, ρ, φ and γ in the formulae are constant in each of the problems.

To assess the accuracy of the calculated method, it is convenient to assess the coefficient of variation V_k of the value to be calculated. To calculate the coefficient of variation of the function from the coefficient of variation or standard deviations of random arguments, an expression known from error theory may be used.

If $k = f(x,y,z)$,

$$V_k = \frac{1}{\sqrt{(n)}} \sqrt{\left[\left(\frac{\overline{x}}{\overline{k}} \frac{\partial k}{\partial x} V_x\right)^2 + \left(\frac{\overline{y}}{\overline{k}} \frac{\partial k}{\partial y} V_y\right)^2 + \left(\frac{\overline{z}}{\overline{k}} \frac{\partial k}{\partial z} V_z\right)^2\right]} \tag{10.41}$$

$$V_k = \frac{1}{k\sqrt{(n)}} \sqrt{\left[\left(\frac{\partial k}{\partial x} \sigma_x\right)^2 + \left(\frac{\partial k}{\partial y} \sigma_y\right)^2 + \left(\frac{\partial k}{\partial z} \sigma_z\right)^2\right]} \tag{10.42}$$

where

x, y, z are random arguments;
$\sigma_x, \sigma_y, \sigma_z$ are the root-mean-square deviations;
$\bar{x}, \bar{y}, \bar{z}$ are their average values;
V_x, V_y, V_z are their coefficients of variation.

Error in calculating the permeability coefficient for a rock mass with filled fractures

Now an attempt is made to calculate the error of the permeability coefficient determined by the tensor method from expression (10.3). As applied to the designation in (10.3) the expression for calculating the coefficient of variation of the function is in the following form:

$$V_k = \frac{1}{\sqrt{n \cdot 2}} \sqrt{\left\{ \sum_{i=1}^{r} \left[\left(\frac{\bar{b}}{k} \frac{\partial k}{\partial b} V_b \right)^2 + \left(\frac{\bar{a}}{k} \frac{\partial k}{\partial a} V_a \right)^2 + \left(\frac{\bar{k}_{fr}}{k} \frac{\partial k}{\partial k_{fr}} V k_{fr} \right)^2 \right.\right.}$$
$$\left.\left. + \left(\frac{\bar{\alpha}}{k} \frac{\partial k}{\partial \alpha} V_\alpha \right)^2 + \left(\frac{\bar{\beta}}{k} \frac{\partial k}{\partial \beta} V_\beta \right)^2 \right]_i \right\}$$

(10.43)

where

n is the number of fractures in a set;
r is number of fracture sets;
a, b, k_{fr}, α and β are random values.

For a rock mass with filled fractures (also from formula (10.3)):

$$\frac{\partial k}{\partial b} = \sum_{i=1}^{r} \cdot \frac{k_{fr}}{a_i} A$$

$$A = \{1 - [\sin\beta_i(\cos\alpha_i\cos\varphi + \sin\alpha_i\cos\rho) + \cos\beta_i\cos\gamma]^2\}$$

$$\frac{\partial k}{\partial a} = -\sum_{i=1}^{r} \frac{b_i k_{fr}}{a_i^2} A$$

$$\frac{\partial k}{\partial k_{fr}} = \sum_{i=1}^{r} \frac{b_i}{a_i} A$$

$$\frac{\partial k}{\partial \beta} = -\sum_{i=1}^{r} \frac{b_i k_{fr}}{a_i} \{2(\cos\alpha_i \cos\varphi + \sin\alpha_i \cos\rho) \times \cos\gamma \cos2\beta_i$$

$$+ \sin2\beta_i([\cos\alpha_i \cos\varphi + \sin\alpha_i \cos\rho]^2 - \cos^2\gamma)\}$$

(10.44)

Example

Suppose a rock mass has one set of fractures with stable parameters along which the permeability coefficient is to be determined. The number of measurements $n = 25$. The measurements are taken with errors typical of the above methods (see Section 5.3): coefficient of variation $V_b = 0.3$, $V_{kfr} = 0.5$ and $V_a = 0.1$. Such coefficients of variations characterize high errors in measuring fracture parameters. The average values: $\bar{b} = 0.001$ m, $a = 0.1$ m, $k_{fr} = 100$ m/day, $\alpha = 0$ and $\beta = 0$.

Calculating from (10.43) and (10.44), the coefficient of variation is obtained of the sought value of the permeability coefficient (12%). The error in measuring the permeability coefficient of a fracture filler reaches $V_{kfr} = 1$ thus allowing for a two- and three-fold difference between the measured and actual values, whereas the error of the calculated value increases up to 21%, other conditions being equal. Hence, the result of the calculation points to the order of the magnitude of the rock mass permeability coefficient, which can often be sufficient to solve the problem.

Thus, when calculating permeability of a rock mass with filled fractures, the errors in measuring fracture width, spacing and filler permeability coefficient affect the result, whereas the errors in measuring fracture orientation by the existing field methods are minor and do not affect the final results.

The natural variations in fracture parameters are greater than measuring errors.

Suppose in another example average parameters of a fracture system are: $a = 0.1$ m, $b = 0.001$ m, $k_{fr} = 100$ m/day, $\alpha = 45°$, $\beta = 45°$.

New coefficients of variation are taken in accordance with natural distribution dispersion: $V_a = V_b = V_{kfr} = 1.0$, $V_\alpha = V_\beta = 0.3$. The number of measurements in a system is the same ($n = 25$). Using the formula (10.43) the coefficient of variation can be determined of the sought value of the permeability coefficient of about 36%. With the number of measurements increasing to $n = 100$, the error is reduced by one-half to 18%. As in the preceding example, variations in fracture width, spacing and permeabiity coefficient make a major contribution to the total error.

Error in calculating the permeability coefficient for rock mass with open fractures

In a rock mass with open fractures the permeability coefficient is assessed from the formula (10.3). There is no filler permeability coefficient. Derivatives with respect to other fracture parameters required for calculations from (10.43) take the form

$$\frac{\partial k}{\partial b} = \sum_{i=1}^{r} \frac{g \cdot 3b_i^2}{12va_i} A$$

$$\frac{\partial k}{\partial a} = -\sum_{i=1}^{r} \frac{g \cdot b_i^3}{12va_i^2} A$$

$$\frac{\partial k}{\partial \beta} = -\sum_{i=1}^{r} \frac{gb_i^3}{12va_i} \{2(\cos\alpha_i \cos\varphi + \sin\alpha_i \cos\rho)\cos\alpha \, \cos2\beta_i$$

$$+ \sin2\beta_i[(\cos\alpha_i \cos\varphi + \sin\alpha_i \cos\rho)^2 - \cos^2\gamma]\} \tag{10.45}$$

Example

Let the permeability coefficient of one fracture set be assessed. The total number of measurements is $n = 25$. The average fracture spacing is $a = 0.1$ m and the average fracture opening is $b = 0.001$ m. The respective coefficients of variation $V_a = 0.1$ and $V_b = 0.3$, i.e. the distribution parameters are similar to those in the first example yet with no fracture filler. After calculation, the coefficient of variation of the calculated value of the function is approximately 18%. This result is 1.5 times as great as the one for the same conditions provided the fractures have a filler ($V_{kfr} = 0.5$). The coefficient of variation of fracture spacing is fairly large. Yet, it does not make a tangible contribution to the total error. If the accuracy of measuring fracture spacing increases two-fold, namely assuming $V_a = 0.05$, the accuracy of assessing the permeability coefficient will not be enhanced.

Now consider the effect of the fracture orientation measuring error on the accuracy of calculating the permeability coefficient. Assume $V_\alpha = V_\beta = 0.03$ and $\alpha = 45° = 0.8$ rad, all the parameters of the fracture set and measuring errors remaining the same.

Thus, the actual measuring errors of orientation and spacing of open fractures do not practically affect the accuracy of assessing the rock mass permeability coefficient. The accuracy of the result is determined solely by the accuracy of measuring fracture width.

When analysing actual errors in measuring fracture width, it was noted that the coefficient of variation of 30% is observed in the event of fractures whose width measures fractions of a millimetre. Such fractures feature fairly low permeability. In fractures about 1 mm wide the measuring error is smaller. Roughly it may be taken as $V_b = 0.1$. Such are the contributions of the errors to the error of calculating fracture permeability coefficient. With the number of measurements in a set $n = 25$, they are minor except for the contribution of the fracture width measuring error. Yet this error, too, enables the permeability coefficient to be assessed to a high degree of accuracy on the basis of fracture parameters.

When studying fracture systems with considerable natural heterogeneities, the assessment accuracy has already been established for fractures with a filler. For systems with open fractures it may be clear from the following example $a = 0.1$ m, $b = 0.001$ m, $\alpha = 0$, $\beta = 0$, $V_a = V_b = 1.0$, $V_\alpha = V_\beta = 0.3$, $n = 25$; after calculation, $V_k = 0.68$.

For a rock mass with filled fractures the error was twice as small under the same conditions. In this case fracture width variation mostly contributes to the total error. The accuracy of assessing the permeability coefficient decreases several times due to natural heterogeneity. In field experiments the error reflecting natural heterogeneity is also several times

the measuring accuracy. The accuracy may be increased with a larger
number of fracture width measurements, i.e. through the rock mass area
subjected to measurements.

The random error described herein has nothing to do with a systematic
error which cannot be reduced through a larger number of measurements.

Systematic errors in calculating the permeability coefficient by the tensor method

Systematic errors arise from partial discrepancy between the calculated
scheme and the natural process reflected by the said scheme. When using
the tensor method, such a discrepancy may take place in the event of great
dispersion of fracture openings. A model of the tensor permeability theory
presumes that all the fractures of a set are similar in width. It is different
with natural fractures. The application of the tensor method to the natural
environment results in a serious systematic error. The magnitude of the
error will be assessed below.

Another assumption of the tensor theory, which is often disregarded in
practice, is that fractures are infinite and the flow running through them is
independent. Actually, fractures are restricted in length, and the cross-
flow from one fracture to another, namely flow interrelationship, has a
great effect on rock mass permeability.

Consider an error arising from different fracture opening, and the
relation between water flow rate and fracture opening (10.32). The
relation is non-linear. Consider now the flow rate in a system of parallel
fractures differing in width (Figure 10.11(a)). Let water flow through
fractures from boundary A to boundary B. The total water flow passed
through fractures is

$$Q = \sum_{i=1}^{n} q_i = C \sum_{i=1}^{n} b_i^3 \qquad (10.46)$$

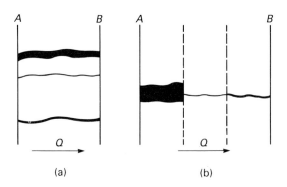

(a)   (b)

Figure 10.11 Parallel (a) and successive (b) arrangement of fractures in rock mass

where b_i is width of every ith fracture in the set, and

$$C = \frac{I_g}{12v} \tag{10.47}$$

Now substitute

$$n \left(\sum_{i=1}^{n} \frac{b_i}{n} \right)^3 \text{ for } \sum_{i=1}^{n} b_i^3$$

and write:

$$Q = C_n \left(\sum_{i=1}^{n} \frac{b_i}{n} \right)^3 \tag{10.48}$$

This substitution is valid in case

$$\sum_{i=1}^{n} b_i^3 = n \left(\sum_{i=1}^{n} \frac{b_i}{n} \right)^3 \tag{10.49}$$

This condition can be met if b_i = const for a system/agregate of fractures, i.e. the average width of a system/aggregate of rock mass fractures is

constant. In this case $\sum_{i=1}^{n} b_i = nb_i$. If the equality is not true.

$$\sum_{i=1}^{n} b_i^3 > n \left(\sum_{i=1}^{n} \frac{b_i}{n} \right)^3 \tag{10.50}$$

Using simple numerical examples, it can be shown to which substantial error this might lead when calculating the total flow passed through fractures (Table 10.5).

As can be seen, the presence of one fracture whose width is an order of magnitude greater than that of 100 other thinner fractures may result in a considerable (three-fold) error in calculating flow rate. Should the width of the above fracture be two orders of magnitude greater than that of other parallel fractures, the flow rate calculated from average fracture opening will be 1200 times lower than the actual one. Such are errors arising from replacing a random argument by its mathematical expectation. They are especially high for left-asymmetric width distribution, in particular log-normal distribution. In the event of right asymmetry of width distribution which cannot be actually observed under natural conditions the error in calculating flow rate would not be so great. So if one fracture width is an order of magnitude smaller than that of 100 other wider fractures, the flow rate calculated from average fracture opening will be only 1.02 times lower than the actual one. Theoretically it is then possible to measure only wide fractures in the field and ignore thin ones.

Table 10.5 Numerical examples

Model parameters				$\sum\limits_{i=1}^{n} b_i^3$	$n\left(\dfrac{\sum\limits_{i=1}^{n} b_i}{n}\right)^3$	$\dfrac{\sum\limits_{i=1}^{n} b_i^3}{n\left(\dfrac{\sum\limits_{i=1}^{n} b_i}{n}\right)^3}$
fracture opening (cm)		number of fractures with openings b_1 b_2				
b_1	b_2	n_1	n_2			
0.01	0.10	1	1	1.001×10^{-3}	0.333×10^{-3}	3.0
0.01	0.10	100	1	1.1×10^{-3}	0.133×10^{-3}	8.5
0.01	0.10	1	100	1.0	0.098	1.02
0.01	0.10	1	1	1.1	1.0	1.1
0.01	0.10	10^6	10^6	1000.0	1000.0	1.0
1.0	10.0	1	1	1001.0	333.0	3.0
0.01	1.0	1	1	1.0	0.254	4.0
0.01	1.0	100	1	1.0001	8.1×10^{-4}	1200.0
0.01	1.0	100	10	10.0001	0.11	91.0

In the event of successive arrangement of fractures with different opening (Figure 10.11(b)), calculations based on average width also result in a considerable error. To assess its magnitude, refer to a formula known from the seepage theory for the average permeability coefficient of a bedded formation in the direction perpendicular to bedding. Assume that fractures determine permeability coefficients of beds separated with dashed lines. The permeability coefficient of the member between the boundaries A and B can be calculated. When calculating from average fracture opening, it proves to be much higher than when doing this from individual values. For example, let two fractures be arranged in succession. Their opening is $b_1 = 10$ cm and $b_2 = 1$ cm. Fracture lengths are equal ($l_1 = l_2 = 100$ cm). It is required to determine flow rate on the basis of average and individual fracture opening. The permeability coefficient of the fractures arranged in succession can be calculated using formula (10.51) through individual values of fracture opening:

$$K = \frac{l_1 + l_2}{(l_1/C \cdot b_1^3) + (l_2/C \cdot b_2^3)} \tag{10.51}$$

where C is a constant determined by viscosity of the liquid. The permeability coefficient equals $2C$. The permeability coefficient calculated from the arithmetic mean of fracture opening is $166.4C$. Hence, when calculating from average fracture opening, the result is about 100 times higher. If one wide fracture is taken per 100 thin fractures, instead of one wide and one thin fracture, the error will be much smaller. Fracture width dispersion in the first example is several times as great as that in the second example. The systematic error increases accordingly. So this type of error is related to fracture width dispersion and always increases as dispersion grows.

Hence, the following conclusion can be drawn. When calculating water flow in fracture systems, the average parameters can be taken into account only in the event of minor variations in fracture width. The presence of some long fractures whose width is one or two orders of magnitude greater than that of most other fractures is typical of zones with hypergene, fault-line and other local transformations of fracturing.

Results of experimental verification of calculation accuracy

The error in calculating the permeability coefficient was defined above as a function of errors in measuring such fracture parameters as width, orientation and spacing. Now another method can be used to assess the same error of the permeability coefficient. The calculated values should be correlated with experimental data. The error is to be found as a deviation of the calculated value from the experimental one.

The difference between the calculated and experimental values should be treated tentatively as an error of the calculated value. This implies that the experimental values are assumed to be true. Actually, the experimental values, much like the calculated ones, are in error. Later, the effect on the result of such an assumption will be assessed, but now assume that the experimental error does not differ from zero. The discrepancy between the experimental and calculated values is to be considered as a magnitude of the error, a random value featuring normal distribution. The standard deviation and mathematical expectation of the random value make up random and systematic errors of the calculated permeability coefficient.

Error calculation

To compare the calculated and experimental data, the results of laboratory studies in flumes, filling of test holes, injections, as well as the results of calculating permeability coefficients of rock masses (see Section 10.4) can be used. These data are tabulated in Table 10.6.

The errors were calculated with the following formulae

$$\Delta_t = \log K_e - \log K_t \tag{10.52}$$

$$\Delta_l = \log K_e - \log K_l \tag{10.53}$$

where

Δ_t and Δ_l are absolute errors in determining permeability coefficient with the tensor method and the method of linear elements respectively;

$\log k_e$, $\log K_t$, $\log K_l$ are logarithms of permeability coefficient determined experimentaliy, with the tensor method and the method of linear elements respectively.

The distribution of errors calculated in Table 10.6 may be approximated by normal curves (Figure 10.12) with the following parameters. For errors of the tensor method the average value is -0.79 and the standard deviation is 0.43. For errors of the method of linear elements the average value is -0.01 and the standard deviation is 0.32.

While comparing these two methods, it may be seen that the tensor method is responsible for a large systematic error. The calculated value of the permeability coefficient proved to be higher than the experimental one in all the cases except one, exceeding the latter, on average, by 6.2 times. However, the correction by dividing the calculated value by the appropriate coefficient cannot lead to accurate assessment since the standard random error of the calculated values makes up 43% of the permeability coefficient logarithm. In the event of the method of linear elements the errors will be smaller, the systematic error being practically nil and the random error making up about 32% of the result in the same expression.

Actually, the calculated errors which are so far attributed to inadequate calculating techniques comprise both calculated and experimental errors. These are all reasons to believe that these two types of error are independent. Hence from the known theorem of the theory of chances:

$$\sigma_{tot} = \sqrt{(\sigma_e^2 + \sigma_t^2)} \tag{10.54}$$

$$\sigma_{tot} = \sqrt{(\sigma_e^2 + \sigma_l^2)} \tag{10.55}$$

where

σ_{tot} is standard deviation of distributions (see Figure 10.12);

$\sigma_e, \sigma_t, \sigma_l$ are all standard errors of the experimental tensor method and method of linear elements respectively.

Table 10.6 Comparison of permeability coefficients calculated from fracture parameters with experimental values

Experiment type	Experiment no.	Experimental results			Calculated permeability coefficients and their deviation from experimental results					
		permeability coefficient (m/day)			tensor method		method of linear elements			
		K	log K	K (m/day)	log K	error Δₜ	K (m/day)	log K	error Δₜ
1	2	3	4	5	6	7	8	9	10
Laboratory	4a	160	2.20	180	2.25	−0.05	55	1.74	−0.46
	4c	30	1.48	365	2.56	−1.08	37	1.57	+0.09
	4d	30	1.48	333	2.48	−1.00	42	1.62	−0.24
	4e	50	1.70	462	2.66	−0.96	113	2.05	−0.35
	5a	172	2.23	931	2.97	−0.54	164	2.22	+0.01
	5b	50	1.70	1211	3.08	−1.38	224	2.35	−0.65
	6a	590	2.77	1916	3.28	−0.51	512	2.71	+0.06
	6b	172	2.23	1712	3.23	−1.00	174	2.24	−0.01
	6c	285	2.45	880	2.94	−0.49	312	2.49	−0.04
	6d	221	2.34	894	2.95	−0.61	176	2.24	+0.10
Filling hole	1501	120	2.08	1389	3.14	−1.06	324	2.51	−0.43
	1502	14	1.15	346	2.54	−1.39	48	1.68	−0.53
	1503	7.0	0.85	64.0	1.94	−1.09	45.0	1.65	−0.80
	1507	6.0	0.80	125.0	2.10	−1.30	20.0	1.30	−0.50
	1509	0.4	−0.40	4.5	0.65	−1.05	0.2	−0.70	0.30
	1505	50.0	1.70	452.0	2.65	−0.95	–	–	–
Injection into hole	518	1.25	0.59	–	–	–	1.0	0.00	0.59
	2011/1	0.22	−0.66	3.5	0.54	−1.20	0.1	−1.00	0.34
	2011/2	0.22	−0.66	0.3	−0.52	−1.18	0.2	−0.70	0.04
	2012/3	0.22	−0.66	1.2	0.08	−0.74	0.5	−0.30	−0.36
	1026/4	0.57	−0.24	5.0	0.70	−0.94	1.4	0.15	−0.39
	2015/7	0.57	−0.24	0.3	−0.52	0.26	0.2	−0.70	0.46
	2015/8	0.57	−0.24	1.5	0.18	−0.06	0.4	−0.40	0.16
	2015/9	0.41	−0.39	1.0	0.00	−0.39	0.2	−0.70	0.31
	2014/9	1.1	0.02	2.6	0.41	−0.39	0.5	−0.30	0.32
	2014/10	1.1	0.02	12.5	1.09	−1.07	–	–	–
	2014/11	0.21	0.67	195.0	2.29	−1.62	3.1	0.49	0.18
	2014/12	0.21	0.67	86.0	1.93	−1.26	5.1	0.70	−0.03

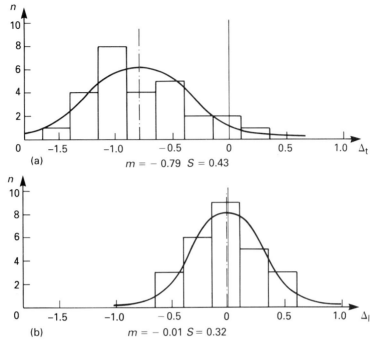

Figure 10.12 Distribution of errors in determining permeability coefficient on the basis of fracture parameters: (a) with tensor method; (b) with method of linear elements

Hence, assuming the experimental error σ_e and knowing σ_{tot}, one may verify errors of the calculated methods. As is known (Chernyshev, 1972), the metrological error in determining specific water absorption in a single injection makes up about 10% of the result at the specified reliability level. The experiments run in flumes (see Section 10.4) give the same metrological error. This being the case, the error of the tensor method will go down to 41% according to formula (10.54) and that of the method of linear elements will go down to 30% according to formula (10.55).

Thus, the established calculated errors are fairly high. Are they commensurable with errors allowed for in isolated injections and pumpings? The errors in testing individual holes are also significant. They are caused, among others, by such factors as colmatage, suffosion, and formation fracturing at high pressures.

In summary, the following conclusions can be drawn:

1. Geologists have fairly accurate methods for measuring rock fracture parameters which enable them to calculate the permeability coefficient with metrological accuracy at the level of direct field determination of the permeability coefficient with test injections and pumpings. Further improvements in the methods for field fracture measurements lie in computerization. The accuracy of measuring fracture orientation may be reduced, that of measuring fracture spacing may be kept at the

present level, and that of measuring fracture opening should be increased.

2. Natural heterogeneity of fracture systems results in a large error of analogy which exceeds a metrological error by several orders.

3. While comparing the tensor method and the method of linear elements, it may be seen that the latter is more accurate. Permeability coefficients calculated by the method of linear elements have no systematic error. As for a random error, it is 1.5 times as small as that when calculating with the tensor method.

10.6 Determination of water permeability of fractured rock masses using the method of finite elements*

Permeability coefficients were determined for a mass of fractured soft rocks composing the flanks of the water reservoir of a hydraulic project under construction. The area in question was cut by an adit in the left-bank and consists of clay marls, dolomites and limestones. The fracturing diagram (Figure 10.13) was drawn up on the basis of graphic representation of fractures on the adit wall. Fractures are filled in with ice, a loose mineral filler being practically absent.

To determine permeability coefficients of the fractured rock mass, an area has been outlined in the form of a circle. This is done purposely so as to determine permeability coefficients not only in some specified direction, but in different directions covering a sector from $-90°$ to $90°$. With this in mind there are zones of inlet and outlet of the seepage flow and two opposing impervious boundaries. The position of these zones is shown in Figure 10.13, the hydraulic gradient vector being directed horizontally. Water permeability is calculated in several stages, each one being characterized by a particular position of pervious and impervious boundaries which are rotated counterclockwise in an automatic mode.

Water permeability was studied with the following variation in the hydraulic gradient $0.1 \leqslant J \leqslant 10.0$. It was assumed that if the Reynolds number Re exceeded the critical values Re_{cr} in the fracture elements, the relationship between flow and hydraulic gradient would be described by the principles governing turbulent flow:

$q = 4.7b \sqrt{[(g^4/\nu)b^5 J^4]}$ in smooth open fractures
$q = bk_{fr} \sqrt{(J)}$ in fractures with a filler

There is a great variety of data obtained following the calculations, only some of which deserve special attention.

Figure 10.14(a) shows a relationship between total flow running through the rock mass and hydraulic gradient. The relationship is non-linear. It is only in the initial portion ($J < 0.15$) that the linear flow–gradient relationship is observed. This, as is evidenced by the detailed analysis of the calculation results, corresponds to the range of gradients featuring laminar flow of water through all the fracture elements. The foregoing is confirmed

---

* Written by V. V. Semyonov and N. N. Shevarina.

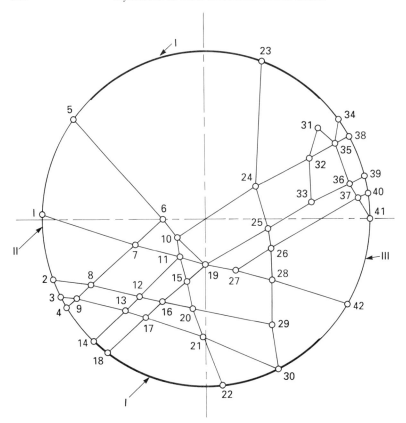

Figure 10.13 Fracturing diagram: I – impervious portions; II, III – seepage flow inlet and outlet boundaries; 1, 2, . . . , 42 – node numbers

by the graph (Figure 10.14(b)) showing the integral permeability coefficient k as a function of log J, wherein can be seen an area with a constant permeability coefficient at log $J < -0.82$ ($J < 0.15$). Should the hydraulic gradient increase, flow turbulence takes place, thereby reducing the integral permeability of the rock mass.

The diagram of permeability coefficients calculated at hydraulic gradient $J = 0.1$, 0.5, 5 and 10 are shown in Figure 10.15. The relationships between permeability coefficients and direction point to a large permeability anisotropy in the rock mass. The form of the diagram is in good agreement with fracture system features. Large values of the permeability coefficient correspond to a sector wherein a fracture with the width b of 0.3 cm bounded by nodes 1 and 42 (Figure 10.13) terminates in pervious portions of the contour. The permeability coefficient reaches a maximum $k = 558$ m³/day when a fracture with width b of 0.2 cm bounded by nodes 2 and 30 also terminates in the pervious boundaries. In other portions of the diagram the permeability coefficients are by several orders of magnitude smaller, and permeability is mostly determined by water flow running through narrow fractures ($b = 0.005$ cm).

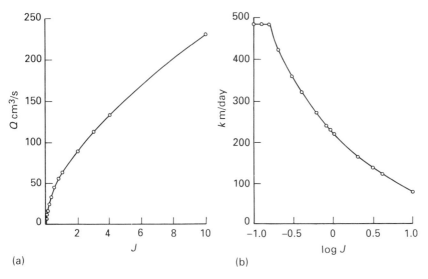

(a)   (b)

Figure 10.14 Relationships between seepage flow rate Q (a) and integral permeability coefficient k (b) and hydraulic gradient J

The diagrams of permeability coefficients thus obtained were processed by the method of least squares to determine the position of principal axes of permeability anisotropy and describe the permeability coefficient–direction relationship. Interpretation of the circular diagram as an ellipse of permeability anisotropy at $J = 0.1$ is shown in Figure 10.15. The authors

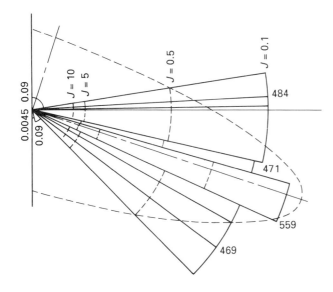

Figure 10.15 Circular diagram of integral permeability coefficients k, m/day

are well aware that the interpretation of the diagram as an ellipse is far from being optimum, but think it can be given as an example of a general methodological approach to the interpretation of results. The ellipse has the following parameters: permeability coefficient is 556 m³/day in the direction of the major axis forming an angle of $-17.8°$ with the horizontal and 155.1 m³/day in the direction of the minor axis.

The results obtained are of great importance to the realization of numerical solutions of seepage problems in fractured rock masses. They help to ascertain relationships between permeability coefficients and hydraulic gradient and introduce them as a variable parameter into Darcy's law without any attempt to describe features governing water permeability of some fractured medium in transient and turbulent flow conditions.

The reliability of the above-mentioned mathematical method for modelling seepage in fracture systems is confirmed by the comparison of the results of numerical investigations with data obtained from physical modelling of groundwater flow through a fractured rock mass (Section 10.4). In particular, the integral flow rate Q was determined for a system of water-conducting fractures shown in Figure 10.7(b) (sample 4d) at a level drop (head pressure difference) $\Delta h = 6$ cm and a temperature of $t° = 20$ °C. The calculated flow rate was 1.425 cm³/s, which agrees well with the experimental relation $Q\ (\Delta h)$ shown in the above diagram.

10.7   Example of the determination of rock mass permeability coefficients and water flow velocities in fractures

Initial data for calculation

The rock mass of the right-bank flank of the dam whose permeability is to be determined is composed of clastic and carbonate sedimentary rocks of the Upper Cambrian (Figure 10.16). There are two suites, one of which, 450–500 m thick, comprises sandstones, siltstones and argillites with interstitial carbonate cement. There are also dolomites and limestones. The other suite is composed of sandstones, fine-grained cavernous limestones, dolomites with interlayers of marls, siltstones and argillites up to 10 cm thick. The rocks are tough and feature clearly defined systematic fracturing. The thickness is about 50 m.

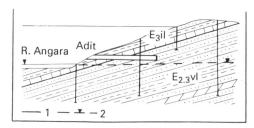

Figure 10.16 Schematic geological section, right-bank flank of dam 1 – stratigraphical boundary; 2 – groundwater level

Bedding is conformable and evenly dipping at an angle of 10–12° south-south-eastwards, i.e. in the direction of the river.

Rock fracturing

Planetary fractures are most common. Crush zones 1–2 to 15–20 m thick were intersected by boreholes. Gravity deformations were responsible for major (from the viewpoint of engineering hydrogeology) transformations in the fracture system in the rock mass under study. In the near-surface layer the width of gravity fissures is by two to three orders of magnitude greater than that of fractures located at depth. The length of large gravity fissures exceeds that of major fractures by 10–100 times. The gravity fissures form a sparse special system of fractures. Calculations deal individually with rock mass permeability through fractures and with that through gravity fissures.

For calculations using the method of linear elements, graphic representation of fractures was made (Figure 10.17) in individual sections of the adit wherein a persistent fracture system was encountered. No calculations were made for the areas featuring a non-persistent system of fractures: the permeability coefficient of rocks in a block does not differ from zero. Hence the rock mass is impervious in areas with a non-persistent fracture system. The areas with gaping gravity fissures were not taken into account when calculating the permeability coefficient in individual sections of the adit. Water flow in the system of gravity fissures was calculated using the method of linear elements individually, irrespective of the calculations of fracture permeability.

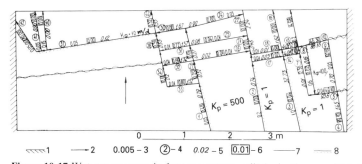

Figure 10.17 Water movement in fracture system, adit 1, depth range 58.0–65.3 m: 1 – impervious rock mass boundary; 2 – flow direction; 3 – fracture opening, cm; 4 – fracture number; 5 – fracture hydraulic gradient; 6 – fracture flow velocity; 7 – active open fracture; 8 – active filled fracture

Based on the results of the experiments with isolated fractures, roughness in such wide fractures as those in the rock mass under study does not greatly contribute to hydraulic resistance. The error in measuring fracture width in the field is so large that it masks the error due to ignoring roughness when calculating the permeability coefficient. Therefore for calculations, fractures are assumed to be smooth.

A small number of fractures in the rock mass have a loose clastic filler consisting of products of weathering or tectonic crushing. Fractures filled with mylonite were not taken into account as they are impervious. Such fractures are known not only to block the passage of water, but even to retain the flow running through them in a crosswise direction.

Initial data for calculating the permeability coefficient of a fissure set with the method of linear elements

The calculation of water flow in large fractures genetically related to gravity deformations was made on the basis of the diagram shown in Figure 10.18 and the relevant table (Table 10.7) listing the parameters of all fractures shown in the diagram.

As is evident from Figure 10.18, fissures form the same two sets which were observed in the adit. The azimuthal dip is 70 and 336°. On the whole, the fissure system provides a continuous passage for water flow from the upper to the lower pond. Many long fissures feature different opening

Table 10.7 Main calculated parameters of fracture system

Fracture element no.	Element bounding node no. (Fig.10.18)	Length (cm)	Width (cm)	Fracture element no.	Element bounding node no. (Fig.10.18)	Length (cm)	Width (cm)
1	1.7	84	0.3	31	4.9	15	0.3
2	7.11	37	0.3	32	9.8	15	0.3
3	11.19	22	0.3	33	8.7	48	0.3
4	19.27	25	0.3	34	7.6	30	0.3
5	27.28	30	0.3	35	14.13	37	0.005
6	28.42	65	0.3	36	13.12	16	0.005
7	2.8	34	0.2	37	12.11	45	0.005
8	8.12	40	0.2	38	10.24	75	0.3
9	12.16	19	0.2	39	24.32	50	0.3
10	16.20	25	0.2	40	32.35	23	0.3
11	20.29	66	0.2	41	35.38	14	0.3
12	3.9	18	0.2	42	18.17	44	0.005
13	9.13	38	0.2	43	17.16	18	0.005
14	13.17	18	0.2	44	16.15	25	0.005
15	17.21	50	0.2	45	15.19	20	0.005
16	21.30	65	0.2	46	19.25	58	0.005
17	5.6	112	0.005	47	25.33	40	0.005
18	6.10	17	0.005	48	33.36	35	0.005
19	10.19	32	0.005	49	36.39	14	0.005
20	10.11	15	0.005	50	27.26	32	0.005
21	11.15	20	0.005	51	26.37	84	0.005
22	15.20	22	0.005	52	37.40	9	0.005
23	20.21	24	0.005	53	31.35	19	0.005
24	21.22	44	0.005	54	35.36	33	0.005
25	23.24	104	0.005	55	36.37	14	0.005
26	24.25	34	0.005	56	37.41	18	0.005
27	25.26	16	0.005	57	31.32	25	1.55
28	26.28	24	0.005	58	32.33	35	1.55
29	28.29	35	0.005	59	34.35	22	1.55
30	29.30	36	0.005				

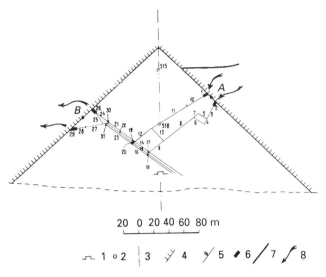

20 0 20 40 60 80 m

⌐ 1 ○ 2 | 3 ⤢ 4 ⤡ 5 ◆ 6 ╱ 7 ⤢ 8

Figure 10.18 Arrangement of gravity fissures under a dam, Boguchanskaya Hydro, Angara: 1 – adit mouth; 2 – boreholes; 3 – dam axis; 4 – dam contour; 5 – unfilled fracture portion and its number; 6 – fracture portion filled in with loose material; 7 – reservoir bank; 8 – inflowing and outflowing water

along their length. Occasionally they are filled in with some loose clastic material. To allow for variations in fracture opening lengthwise, some fissures were broken into segments. Fracture opening and permeability coefficient of a filler, if any, were assumed to be constant within each particular segment. The numerical values of fracture opening were taken mostly from measurements made in the adit. Thus, the model reflects the rock mass conditions at a depth of 10–30 m. It is only at the inlet and outlet (*A* and *B* in Figure 10.18) that fracture opening was taken from geological survey data. This reflects the conditions of flow entry into the rock mass in the upper pond through the surficial belt and appropriate conditions of flow discharge in the lower pond.

Results of calculating the permeability coefficients and flow velocities in fractures

Application of the method of linear elements to the fracture system

The rock mass permeability coefficient was calculated by the method of linear elements for individual sections of the adit with a clearly defined fracture system (Table 10.8). Heterogeneity of the results obtained is very similar to that of the injection results. Strictly speaking, the values obtained cannot be treated as a permeability coefficient, a constant characterizing the rock mass. The calculated value is determined not only by rock mass structure, but also by flow gradient. The gradients used in calculations lead to turbulent flow in fractures of such zones as an interval

Table 10.8 Rock mass permeability coefficients in different adit intervals (adit no.1) (m/day)

Interval (m)	Graphic representation of adit	Direction	Permeability coefficients at flow gradients	
			1.0	0.4
5.0–11.0	Left wall	North	9.0×10^2	1.3×10^3
30.0–37.0	Left wall	Vertical	4.3×10^2	6.2×10^2
30.0–37.0	Left wall	North	0.03	0.03
30.5–41.0	Roof	East	510.0	760.0
30.3–41.0	Roof	North	9.5	9.0
35.6–41.2	Right wall	Vertical	725.0	1054.0
35.6–41.2	Right wall	North	0.5	0.5
53.6–61.0	Roof	North	35.0	35.0
53.6–64.0	Roof	East	180.0	260.0
53.6–64.0	Roof	North	0.7	0.7
58.0–65.3	Right wall	Vertical	1.2	1.2
58.0–65.3	Right wall	North	4×10^{-4}	4×10^{-4}

of 5–11 m. The flow rate–head pressure relationship becomes non-linear, and the permeability coefficient becomes variable. The effect of flow turbulence on the calculation results is obvious from the table as regards a number of zones with high water permeability.

While comparing the results of calculations for different directions, it may be seen that water permeability in all the adit intervals in the lengthwise direction (northwards) is much lower than that in the crosswise direction (eastwards). Such a relation between the permeability coefficients agrees with the actual geological situation. Water permeability along the release fractures (across the adit) is apparently higher than that across them. On the whole, the ellipsoid of permeability anisotropy of the rock mass in the right-bank flank of the dam is apparently an ellipsoid of revolution, with a short axis positioned horizontally across the river. The ratio between the short and long axes is 0.01 or 0.1. The degree of anisotropy apparently reduces as we move from the ground surface downwards into the rock mass.

Apart from directional characteristics of permeability, the calculations enable assessment of suffosive resistance of rocks. With the hydraulic slope of the flow $I = 1$, velocities in some fractures reach 200–300 cm/s and individual hydraulic slopes reach 2 or 3 and even 10. With such gradients in individual fractures, a filler and sometimes the fracture wall rocks may undergo washout.

Application of the method of linear elements to the fissure system

It is assumed that fractures take in water at the reservoir water level in the upper pond and discharge it similarly in the lower pond.

The calculations have shown that the rock mass in the right-bank flank of the dam features high water permeability. In the zone of weathering and active gravity deformations over a depth range of 1–15 m, the permeability coefficients are still higher, reaching a few hundred metres per day.

Table 10.9 Results of calculating rock mass permeability coefficient by the tensor method (directions: eastward, northward and towards the zenith)

Adit interval (m)	Permeability coefficient (m/day)								
	calculated without isolation of fractures and fissures			calculated from fracture set			calculated from fissure set		
	east	north	zenith	east	north	zenith	east	north	zenith
0–22.6	5×10^6	23×10^6	27×10^6	3×10^5	9×10^5	12×10^5	7×10^7	35×10^7	42×10^7
23.8–51.1	8×10^4	7×10^4	15×10^4	2×10^3	1×10^3	2×10^3	6×10^5	12×10^5	18×10^5
51.1–83.1	1.5×10^3	6×10^3	7×10^3	5×10^2	2×10^2	6×10^2	1×10^3	50×10^3	60×10^3

The calculation of rock mass permeability coefficient with the tensor method on the basis of the same fracture parameters has led to results that are far from the experimental data, and are untrue. Table 10.9 serves to illustrate the systematic errors of the method.

No significant discrepancy is found between the calculated and experimental data when the method of linear elements is employed for calculations. The example shows that by calculations, water permeability of large areas that cannot be tested with water filling or injection can be assessed. At the same time it has been revealed that permeability coefficients can also be obtained by calculations for such areas where hydrodynamic investigations are impossible due to extremely high water absorption.

10.8   Summary

The geological methods for studying rock fractures, the hydraulics of groundwater movement in rock fractures and the potentialities of computers enable an easy assessment of the rock mass permeability coefficient, flow velocity, gradient and Reynolds number. This in turn enables modelling of permeability testing to define seepage properties of such rock masses which cannot for some reason be tested through pumping and injection. There are numerous cases where direct hydrodynamic testing is impossible. This applies first and foremost to permafrost and dry rock masses, especially those with wide fractures.

The permeability coefficients calculated as a result of modelling are always related to a particular direction. Thus one can assess not only the absolute value of permeability, but also rock mass anisotropy.

The experimental practices and theoretical analysis have shown that the permeability coefficient cannot so far be calculated from fracture parameters to a high degree of accuracy. This largely stems from the fairly low accuracy of fracture width measurements in engineering surveys.

Yet the accuracy of calculations is determined not only by that of initial data, but also by compliance of the model with the natural process. There are two optimum methods employed for mathematical description of flow in fractures and fracture systems, namely the tensor method and the method of linear elements. Comparison under laboratory and field conditions indicates that each of them offers some advantage. The solution obtained on the basis of the tensor method is simpler and more convenient for practical use. However, in cases where fracture parameters such as width, length, and so on, show a great dispersion, which is quite common, the method lacks accuracy. The calculated value exhibits large random and systematic errors. In the event of minor dispersion of fracture opening the tensor method is satisfactory, i.e. its accuracy is quite fair. The method of linear elements is more versatile. It enables the permeability coefficient to be calculated with fair accuracy under diverse natural conditions.

The analysis of water flow in fractures on the basis of the theory of linear elements has considerable promise as regards forecasting suffosion and grouting or colmatage of a fractured rock mass, and some other processes. Assessment of rock mass permeability properties on the basis of fracturing data is largely determined by the accuracy of results. With this in mind a major effort has to be made towards improving measuring accuracy.

Side-by-side with methodological recommendations, some conclusions may be drawn as to the nature of permeability of fractured space in that part of the hypergenesis zone which is in contact with engineering structures. Following calculations of the permeability coefficient, permeability anisotropy, much like permeability, is shown to vary greatly as the ground surface is approached. The permeability coefficient in the plane tangent to the relief increases at the surface. The permeability coefficient in the direction perpendicular to the said plane either increases at a slower rate or even decreases. Another feature of permeability of a fractured rock mass which has been confirmed here is the localization of flows in a fractured rock mass. Most of the flow passing through the rock mass runs along the chains of wide fractures, which so far has not been reflected in the tensor permeability theory. In a fractured rock mass water flow is generally localized in individual fractured zones and within those zones in large fractures connected with one another.

Chapter 11
Determination of rock mass deformation modulus on the basis of fracture parameters

11.1 Theoretical aspects and practical methods

A rock mass is a complex natural structure, its different parts being subjected to different degrees of compression. Fracture fillers are more compressible than the rock blocks provided the filler is of tectonic or hypergene origin. The hollow space of fractures easily contracts when the rock mass stresses increase. The same structural elements of a rock mass undergo different compression determined by such factors as moisture content and stress. Weathering and technogenic deformations can also affect rock mass compressibility. Thus, rock mass compressibility which is expressed in terms of deformation modulus in engineering calculations depends on many factors, of which the principal ones are: (i) rock mass structural pattern; (ii) deformation modulus of rock mass components; (iii) stressed state at the construction and operating stages; and (iv) rock mass temperature and moisture content at the operating stage.

It is fairly difficult to take due account of all the above factors. Yet wide experience gained in geology, mineralogy and geomechanics permits assessment of the effect of each individual factor.

Deformation of a fractured rock mass has been dealt with in papers by Tarkhov (1940), Walsh (1965), Kandaurov (1966), and Ruppeneit (1975). They treat a rock mass as a two-component medium comprising alternating zones referred to as a 'rock' and a 'fracture'. The deformation modulus of the fracture zone is assessed through the rock deformation modulus. For this a fracture is viewed as a cavity with the walls closing at individual points. One may represent a fracture as two planes having between them springs with deformation characteristics taken for the rock.

More specifically, the contact deformation modulus can be considered as that of minerals coming into contact. Calculations of average mineral deformation moduli have been made by Belikov (1964).

The basic equation of the elastic deformation of a two-component medium in the direction perpendicular to the fracture is as follows:

$$\frac{(a + b)\,d\sigma}{E_1} = \frac{a\,d\sigma}{E_r} + \frac{b\,d\sigma}{\xi E_r} \tag{11.1}$$

According to Hooke's law, deformation of the body consisting of fractures and rock with deformation modulus E_1 is the total of rock deformations within length a with rock deformation modulus E_r and fracture deformations within length b with fracture zone deformation modulus ξE_r; σ is normal stress acting perpendicular to the fracture; E_1 is deformation modulus for the direction perpendicular to the fracture; ξ is mineral contact area on the fracture wall. Reducing the equation to the differential form, and assuming that Hooke's law holds true for the infinitesimal portion of the deformation–stress curve, solution of equation (11.1) will lead to non-linear dependence between deformation and stress which is evidently closer to the observed facts than linear dependence.

From (11.1) the expression for the general rock mass deformation modulus E_1 can be derived (Ruppeneit, 1975). Writing $\eta = b/\xi a$,

$$E_1 = E_r/(1 + \eta) \tag{11.2}$$

For the rock mass with fractures forming different angles with the horizontal plane the general rock mass deformation modulus in the vertical direction is expressed in terms of

$$E_1 = E_r / [1 + \sum_1^k \eta_i(1 - \sin^4\beta_i)] \tag{11.3}$$

where

i is fracture or fracture set index;
β is dip angle of the ith fracture or ith fracture set;
η_i is the geometric characteristic of the ith fracture or ith fracture set;
k is the number of fractures or fracture sets in a rock mass.

The expression (11.3) can be used in the calculation of the general rock mass deformation modulus in any direction differing from vertical. In this case β_i is not a dip angle, but an angle between a fracture (fracture set) and the area perpendicular to the direction of the sought E_1.

All the parameters of equation (11.3) can be determined using existing techniques, except for the rock contact area which is governed by such factors as reworking of the fracture surface through geological time, composition of fracture wall minerals, and fracture closing pressure. Yet Ruppeneit suggested that $\xi = 3 \times 10^{-4}$ be taken as a constant.

Attempts were made to determine the relative rock contact area on the basis of fracture surface configuration. Ruppeneit (1975) proceeded from the assumption that the ground surfaces of tectonic fractures have a larger rock contact area than tension fractures. However, no substantial progress was made in this respect. Actually, the rock contact area is determined by pressure and compressibility of material at contact points rather than by surface roughness.

The fracture wall contact area can be determined on the basis of equilibrium between the wall compression pressure and resistance to compression of the fracture wall material throughout the contact area; on the other hand, the stronger the contact rock, the smaller is the contact area at constant pressure. The relative rock contact area can be expressed in a first approximation as

$$\xi = \sigma/R_s \tag{11.4}$$

where

σ is normal stress in the direction perpendicular to the fracture, MPa;
R_s is ultimate rock resistance to compression in the volume characteristic of the individual fracture wall contact, MPa.

Relevant methods are used in mineralogy to determine mineral hardness by the static indentation method. Indenters of different shape (balls, cylinders, cones) are pressed into the rough surface of a mineral. Based on the indentation area, hardness can be determined as a ratio of pressure to indentation area (kg/cm^2). This experience simulates the processes occurring upon fracture closing, i.e. when protrusions of one fracture wall are pressed into the rough surface of the other wall. In both cases this results in elastic deformation, plastic flow and brittle failure being developed in one and the same proportion. At the indenter and test mineral contact, stresses reach tens of tons per square centimetre. Stresses of the same order develop at fracture wall contacts at 10–30 kg/cm^2 external pressure.

Based on the similarity between the indentation experiment and fracture wall contact processes, let R_s be replaced by the hardness H in expression (11.4):

$$\xi = \sigma/H \qquad (11.5)$$

To confirm our assumption, ξ based on (11.5) can be determined for some specific cases. Quartz hardness is known to be ca. 1000 kg/mm^2. In this case the fracture walls made up of quartz grains with a normal rock mass pressure of 30 kg/cm^2 have a contact area 3×10^{-4} of the total fracture area.

Now write equation (11.1) with (11.5):

$$(a + b)\, d\sigma/E_1 = a d\sigma/E_r + bH d\sigma/\sigma E_r \qquad (11.6)$$

where $(a + b)d\sigma$ is deformation of the block-fracture pair.

After integrating in the range from σ_1 to σ_2, an expression is obtained describing non-linear elastic deformation of the two-component medium:

$$(a + b)(\sigma_2 - \sigma_1)/E_1 = a(\sigma_2 - \sigma_1)/E_r + bH\ln(\sigma_2/\sigma_1)/E_r \qquad (11.7)$$

With σ_1 fixed and σ_2 increasing, deformation increment decreases due to fracture closing. Hence one may deduce a formula to assess an increment of the rock mass deformation modulus with the stress ranging from σ_1 to σ_2:

$$E_1 = \frac{(a + b)\,(\sigma_2 - \sigma_1)\,E_r}{a(\sigma_2 - \sigma_1) + bH \ln \sigma_2/\sigma_1} \qquad (11.8)$$

As is evident from expression (11.8), with the stress σ_2 increasing relative to the initial stress σ_1, an increment of the deformation modulus gradually decreases and tends to zero. As this takes place, the rock mass deformation modulus increases and tends to the rock deformation modulus since with ξ increasing, the second member in expression (11.3) tends to zero. Actually, it may reach the latter only if the fracture walls close throughout the surface.

Now use expression (11.5) in the calculation of the rock contact area.

For this it is necessary that the minerals coming into contact on the fracture wall surface, and their microhardness, be determined on the basis of petrographic data. Constant rock petrography along the entire length of a fracture is essential for the problem to be solved.

For monomineralic rocks, say marble, the determination of fracture wall contact minerals presents no problem. The H value can be taken for the only mineral (calcite) and ξ can, therefore, be calculated easily. The problem is much more difficult in cases where polymineralic rocks are involved. It presumably calls for a probabilistic solution to specify minerals which can be encountered with the fracture walls under deformation. Mineral properties, in particular microhardness, are well-known so tabulated data may be used.

In addition to mineral hardness, stresses also enter into expression (11.5). As a first approximation, the depth-dependent stress variations follow the Heim hypothesis which may hold true for the upper tens of metres in a rock mass. Then

$$\xi = \gamma h / H \tag{11.9}$$

where

γ is rock volume weight, g/cm^3;
h is depth, cm;
H is hardness of contact minerals, g/cm^2

Zelensky was the first to suggest that stresses and hardness of material in a fracture zone should be taken into account (Ruppeneit, 1975). According to him, fracture zone deformability is determined by rock Brinell hardness, normal stress on the fracture surface and fracture form. The form in turn is described by many parameters which are not easy to determine in nature. Such an approach borrowed from physical metallurgy has a number of shortcomings, e.g. rock hardness is taken as a constant. Rocks represent polymineral aggregates. In the inner space of a fracture each mineral grain is an individual body. That is why the hardness of rock-forming minerals may be discussed, but not that of rocks. The hardness and stress factors are used below to assess rock contact area. It is next to impossible to determine rock contact area through survey measurements. The proposed simplified approach to the determination of this value looks promising.

The relationship between fracture zone deformation and stress also makes it possible to assess variations in water permeability with rock mass pressure increasing. For this let fracture opening be expressed in terms of stress in the water flow rate formula, then:

$$q_\sigma = \frac{\gamma [b(I - H \ln (\sigma_2/\sigma_1)/E_r)]^3}{12\mu} \tag{11.10}$$

where

σ_1 is normal stress in the direction perpendicular to the fracture at which fracture width (b) is measured;
σ_2 is stress for which flow rate is calculated.

For most common rocks the constant H/E_r is in the range 0.2–0.01.

11.2   Checking the method against the results of field engineering–geological studies

The calculations given below were made with a view to checking the applicability of the proposed method to the determination of rock mass deformation moduli on the basis of fracture parameters. Large-scale experimental tests (Ruppeneit, 1975) provided somewhat ambiguous conclusions and it is inferred that not all fractures visible in the outcrop should be taken into account. The author believes that only large fractures may serve this purpose.

This assumption is not flawless. The length of fractures in exploration openings, especially in holes, cannot actually be assessed. Besides, there might be a paradox if fractures commensurable in length with the structure foundation are dealt with exclusively, as is evident from the paper by Ruppeneit. For instance, there might not be a single fracture commensurable with the size of the foundation in the system of horizontal fractures under the Bratsk Hydro dam. Horizontal fractures in dolerites are known to be short. In this case the foundation deformation modulus should not apparently differ from that of a rock sample. Yet it proves to be five to ten times lower. So actually in calculations all fractures have to be taken into account, irrespective of their size.

The theory was verified in the course of studies made in the areas of the Bratsk and Ust-Ilim Hydros (Tizdel, 1962, 1963; Eidelman, 1968; Bolotina, 1970) where the foundations were studied in detail. Concrete gravity dams more than 100 m high apply a load of up to 30 kg/cm^2 to the foundation. This being the case, even dolerite, which is not easily deformable, becomes subjected to great deformation (70 mm). Fractures were described in detail in adits, outcrops, large-diameter holes, and core samples. The two dams were erected on complex-shaped trap sills. The geology of the Ust-Ilim Hydro was described in Section 4.4. Figures 11.1, 11.2, 11.3 show the results of direct experimental determination of the dolerite deformation modulus by different methods and the results of calculations of the same value based on fracture parameters at constant and variable ξ.

While comparing the calculated and experimental data, the deformation moduli determined with the variable rock contact area are generally closer to experimental data. This supports the determination of the rock contact area on the basis of mineral hardness in fracture walls and pressure normal to the fracture. The proposed method for determining the rock contact area at the uppermost levels may result in the fact that the calculated and experimental data are not in good agreement because of weathering. Iron oxides and other minerals with extremely low hardness are formed in fractures upon weathering.

Consider the variations in the mineral composition of a filler for the zone of weathering in the Ust-Ilim intrusion. It has been established that the volume weight of dolerites in the zone of weathering remains constant since it is only rock and fracture fillers that are subjected to weathering directly along the fracture wall. Calcite and chlorite are replaced in fractures by goethite and hydrogoethite whose hardness in the water-saturated condition is about 20 kgf/mm^2. The hardness increases up to

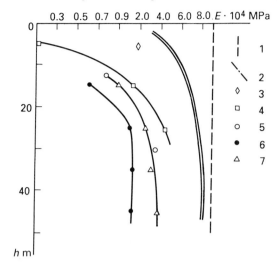

Figure 11.1 Correlation of deformation moduli obtained by different methods for dolerites in the foundation of the Bratsk Hydro dam: 1 – laboratory tests of samples; 2 – pressiometric test for horizontal direction; 3 – measuring foundation compression by means of teletensometers; 4 – plate bearing tests; 5 – back-calculation of modulus based on dam settlement; 6 – calculation from fracture parameters at constant $\xi = 3.10^{-4}$ using formula (11.3) for vertical direction; 7 – calculation from fracture parameters at variable ξ using formulae (11.3) and (11.9)

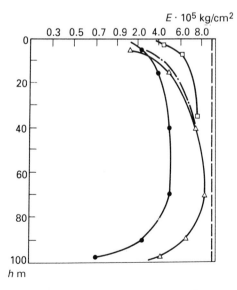

Figure 11.2 Variations in deformation modulus of dolerites (Ust-Ilim intrusion) from top to bottom in slightly fractured zone. For symbols see Figure 11.1. Dot-and-dash line shows a curve plotted with due account taken of fracture-filler weathering

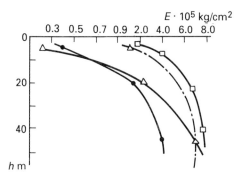

Figure 11.3 Variations in deformation modulus of dolerites (Ust-Ilim intrusion) in the foundation of the power house in zone of increased fracturing and active hypergene transformations. For symbols see Figure 11.1. Dot-and-dash line shows a curve plotted with due account taken of fracture filler weathering.

$100–120$ kgf/mm^2 upon dehydration. Since the rock mass in question is situated under the Angara river bed, the minimum hardness is assumed. The fractures filled with hydrogoethite in the slightly fractured zone at a depth of $0–12.5$ m make up 23% and those filled with calcite and chlorite, 7%. In the zone of increased fracturing, fractures filled with hydrogoethite over a depth range of $0–10$ m account for 100%. Hence the rock contact area in the slightly fractured zone is about $\xi = 3 \times 10^{-4}$ and that in the zone of increased fracturing is $\xi = 7.5 \times 10^{-4}$. Entering the new values of the rock contact area calculated from substituting expression (11.9) into formula (11.2), deformation moduli are obtained of 2.4×10^4 MPa in the slightly fractured zone (Figure 11.2) and 1.2×10^4 MPa in the zone of increased fracturing at a depth of 5 m. These figures are closer to the experimental data than those calculated without regard to weathering, which once more suggests that the rock contact area should be determined using mineral (fracture filler) hardness.

11.3 The effect of rock mineral composition on the fracture wall rock contact area

Up until now discussion has been limited to fractures in whose walls there is either one or several minerals with more or less similar hardness. However, most rocks are polymineral aggregates. In spite of the fact that they are characterized by particular textures and structures, the arrangement of mineral grains is apparently random in a statistical sense.

The problem of determination of rock contact area needs to be considered when the fracture walls contain two minerals whose mutual arrangement is not correlated. These minerals are assigned numbers 1 and 2. Their hardness is H_1 and H_2 respectively. The first mineral occupies S_1 of the fracture area and the second, S_2. It is necessary that the rock contact area of the fracture walls at normal stress σ (Figure 11.4) be determined when $S_1 + S_2 = 1$.

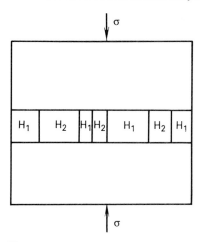

Figure 11.4 Fracture with rock contacts of two types

To solve the problem, all possible mineral combinations at the contacts are examined and a particular hardness is assigned to each mineral combination. While assigning hardness at the contact, it is assumed that grains of the same index coming into contact with each other display their characteristic hardness H_1 or H_2, whereas in the event of grains of different index the hardness of the pair is determined by the yielding grain, i.e. in the pairs 1–2 and 2–1 the hardness H_1 is taken:

Combination	*Hardness*
1–1	$H_1, P_{1-1} = S_1^2$
1–2	$H_1, P_{1-2} = S_1 S_2$
2–2	$H_2, P_{2-2} = S_2^2$
2–1	$H_1, P_{2-1} = S_1 S_2$

The total of probabilities of all the combinations equals unity since the above combinations make up the sum total of events $P_{11}+P_{22}+2P_{12}=1$. The probability of finding hardness H_1 at the contact is

$$P(H_1) = P_{11} + 2P_{12} \tag{11.11}$$

The probability of finding hardness H_2 is

$$P(H_2) = P_{22}^2 \tag{11.12}$$

From expression (11.12) it follows that the probability of finding hardness H_2 at the contact is as low as 1–4%, with hard mineral accounting for 10–20% of the fracture surface. This probability is negligible and can hence be ignored. In this case the effective hardness at the contact is H_1. The sought value of the relative rock contact area can be determined from formula (11.9) through the effective hardness. Therefore, minerals whose total makes up less than 10% of the thin section area can be ignored when calculating the rock contact area. Such common rocks as granodiorite, syenite, gabbro, dolerite can be treated in this problem as monomineral aggregates. The hardness of related rock-forming minerals is not found to vary over wide limits. According to equation (11.12), the hardness of accessory minerals, which may greatly differ from that of rock-forming minerals, may be disregarded.

Consider now the fracture walls as two rigid plates with the fracture space between occupied by contacts of two types with hardness H_1 and hardness H_2. The contacts with hardness H_1 make up A_1 of the total contact area (11.11) and those with hardness A_2 make up A_2 of the total contact area (11.12). The external pressure normal to the fracture is P, kg/cm^2. It gives rise to stress σ_1 at contacts with hardness H_1 and stress σ_2 at contacts with hardness H_2. The respective deformations in the contact areas are ϵ_1 and ϵ_2. The plates are given ideal longitudinal rigidity since the distance between the contacts of different index is incommensurably small as compared with size across the plate. Assume that the normal stress–deformation relationship in the contact zone is linear. Of the above quantities, P, H_1, H_2, A_1 and A_2 are known; σ_1 and σ_2 are to be determined.

A set of six equations in six unknowns can be written:

$$P_1/A_1 = \sigma_1$$
$$P_2/A_2 = \sigma_2$$
$$\sigma_1 = H_1\epsilon_1$$
$$\sigma_2 = H_2\epsilon_2$$
$$P_1 + P_2 = P$$
$$\epsilon_1 = \epsilon_2 = \epsilon$$

These equations express relationships between pressure, area and stress; stress and deformations; pressure at the contacts with H_1 and H_2 and total pressure; equal deformations at all the contacts with no plate buckling. After solving the set of equations:

$$\sigma_1 = P_1H_1/(H_1S_1 + H_2S_2)$$
$$\sigma_2 = P_2H_2/(H_1S_1 + H_2S_2)$$

The rock contact area in this case is equal to the total area of contacts of the first and second kinds:

$$\xi = \sigma_1/H_1 + \sigma_2/H_2$$
$$\xi = P/(H_1S_1 + H_2S_2) \tag{11.13}$$

The expression can be applied to the determination of the contact area in such rocks as granite, gneiss, arkose or marlstone which consist of rock-forming minerals with different hardness.

For polymineral rock the expression may be represented as

$$\xi = \frac{P}{H_1S_1 + H_2S_2 + \ldots + H_nS_n} \tag{11.14}$$

Thus, when making calculations, the rock contact area of the fracture walls should be considered as a variable, much like other fracture parameters in a rock mass. The rock contact area should be determined on the basis of such factors as pressure in the fracture zone, rock mineral composition, mineral hardness, rock mass temperature and moisture content. All visible fractures should be taken into account whose length exceeds that of a rock sample on which the deformation modulus has been determined.

Conclusion

A rock fracture system determines the mechanical properties of a rock mass, provides information on the history of crustal deformations and sometimes is a guide in exploration and estimation of reserves of a wide range of minerals from ore, oil and gas, to building stones. Rock fracture patterns have been studied by geologists for more than 100 years (Pollard and Aydin, 1988). There are fracture classifications. Also described are types of fractures associated genetically with some particular geological processes. In terms of absolute length there are microfractures (less than 0.1 m), macrofractures referred to for clarity as fractures (0.1–100 m) and faults (over 100 m).

A combination of fractures makes up a spatial system which is dense if the spacing between parallel fractures is less than 10 cm. Should the fracture spacing exceed 100 cm a system is considered to be sparse. Among the major characteristics of a fracture system, special mention should be made of persistence governing disintegration of rocks into joint blocks and the mutual arrangement of fractures determining joint block forms. As regards the degree of persistence, a distinction can be drawn between non-, sub- and persistent fracture systems. Jointing in a rock mass with a non-persistent fracture system is fairly imperfect since fractures do not intersect and joint blocks (conventionally termed) are connected through rock bridges. Sub-persistent fracture systems consist of partly intersecting fractures. Jointing in such rock masses is either imperfect or perfect. Persistent fracture systems give rise to fairly perfect jointing, all fractures intersecting and clearly outlining joint blocks. As regards the degree of persistence, three types of fracture system can be roughly treated as three successive stages of fracture system evolution. The first stage is characterized by isolated fractures. Coalescence and intersection of fractures take place at the second stage. The third stage features the process resumed at new levels, with fractures formed inside blocks (original fractures in effusives, feather joints in faults, weathering fractures, etc.) and a system of faults, of which tectonic faults have been studied most comprehensively.

Fracture orientation and angles are determined by the history of rock deformation and expressed as breaking stress tensors. All three possible combinations of principal normal stresses are realized in a rock mass, namely (i) the three principal normal stresses are equal; (ii) two of three principal normal stresses are equal; (iii) all three principal normal stresses are different. Three types of stresses are realized in three types of fracture systems differing in symmetry groups. The fracture system symmetry reflects the stress tensor symmetry. It is expressed statistically in accordance with the statistical character of brittle failure. There are spheroidal, polygonal and regular fracture systems. Particular rock mass anisotropy due to fracturing corresponds to each of the above types of fracture systems. Each geometrical type of fracture system is characterized by a particular genetic type of fracture. The successive superposition of stresses gives rise to an asymmetrical chaotic fracture system eliminating anisotropy.

Rock mass structure and rock properties, along with a stress source, have a great effect on the symmetry of a fracture system. With the formation of a stress field, dissymmetry of the stress source is combined with rock dissymmetry. Hence, the combination of principal normal stresses and related fracture system symmetry depend upon rock bedding and structure.

All genetic types of fractures follow the inheritance principle. A fracture system is characterized by the priority growth of most large fractures which are revived at different stages of evolution of the rock mass. Periodically, the inherited evolutionary trend is terminated under the effect of environmental changes. Newly formed fracture sets appear that are positioned in conformity to the new axes of principal normal stresses. However, with the conditions changed, the old deformation pattern may be regained. Sometimes, a fracture system ceases to exist under metamorphism, hydrothermal healing or weathering. The inherited evolution of fractures then comes to a stop. A superimposed fracture system or one poorly reflecting the previous deformation pattern originates in a rock mass. The old system can be traced as veins and dykes which are not rock mass defects in a mechanical sense and, hence, have little if any effect on the evolution of the superimposed fracture system.

When two growing fractures meet, one of them may cease to grow or the newly formed fracture may be characterized by accelerated evolution depending on the angle of fracture intersection. So there are two heterogeneity levels in a fracture system. The further course of the process leads to new heterogeneity levels. When passing from early to late levels, the number of fractures decreases in a step-like manner and their length increases in a step-like manner.

Statistically inherited fracture evolution is expressed in terms of different distribution of most fracture parameters. Due to the priority growth of large fractures, the distribution of fracture width, length and spacing becomes left asymmetrical and changes from normal to lognormal through geological time. The newly formed fracture sets regularly differ from their old counterparts by smaller length, lower density, as well as by greater variation in all the fracture parameters. This rule does not apply to superimposed fracture patterns.

The mechanism of evolution of inherited fractures becomes very apparent when studying processes of brittle failure, washout and solution of rocks, as well as fracture sealing, which are component parts of the single geological process of fracture evolution. All of them have one major feature in common, i.e. the mathematical functions which describe them are non-linear. Any fracture grows at an accelerated rate under stable geological conditions in accordance with the non-linear law. In other words, of simultaneously growing fractures, large fractures feature more rapid growth. Non-linearity is a fundamental cause of inherited evolution of fracture systems.

The fact that fracture systems evolve in accordance with the inheritance law enables two important practical conclusions to be arrived at. While considering engineering structures as neogenesis in the Earth's crust and applying the inheritance principle to their development under construction and operation, one should (i) allow for possible inherited deformation in a structure in the areas of major rock disturbances; and (ii) try to make the most of the natural structures to organize useful technological processes inherited from nature. The first conclusion has already been put into practice. As for the other, so far it has been used in construction only on a small scale.

The inheritance principle, according to Peive (1956), is a principle of evolution applied to geological processes. Governed by this principle, such processes as rock mass deformation and permeability have been considered with attention concentrated on fractures as elements of weakness in a rock mass. Of these, large fractures were given special consideration since they take the most active part in the above processes. When analysing relationships between seismic dislocations and earthquake parameters, our attention was also focused on the largest fault as a major indicator of earthquake energy.

Engineering–geological studies of fracture systems are aimed at studying each individual fault in detail, assessing macrofractures on a selective basis, and investigating microfractures indirectly on the basis of determination of physico-mechanical properties of rock samples. A quantitative approach is used to represent engineering–geological forecasts in a digital form. Fracturing is characterized by a number of parameters in accordance with the form of models of a fracture medium known from hydrogeomechanics. Both in the field and during desk studies, fracturing is subjected to geological and genetic analysis which is fundamental for extrapolation and interpolation of observational data.

The accuracy of fracture parameter measurements may be satisfactory and high. Standard measuring errors are far smaller than the root-mean-square deviation of natural inhomogeneity. This determines their minor contribution to the total error of calculating rock mass properties on the basis of fracture parameters.

General natural rules governing rock fracture patterns and actual measurements of fracture parameters are widely used in solving a variety of engineering and geological problems. The gap between field and laboratory studies is tending to close gradually (Sergeev, 1974).

In conclusion, forecasting of fracture patterns may be extended to rock masses that have not been examined.

A fracture pattern can be roughly described for some geological body if the following information is available: rock composition and depth of occurrence; in addition, information is required on the related tectonic structure, for example the character of modern tectonic stresses (tension or compression). Unless fractures are studied, a forecast can be made only in a qualitative form.

In making forecasts for sedimentary strata, the following factors are to be taken into account. Fractures in rigid competent layers of the sequence are mostly perpendicular and parallel to the bedding, their spacing being of the order of layer thickness. The extent of the fracture along the layer is generally three to ten times larger than that of the same fracture in the direction perpendicular to the layer. Fractures perpendicular to the layer intersect several layers and terminate at one of the interfaces. In incompetent rocks of the sequence, fractures are much thinner, more closely spaced and feature diverse arrangement. Side-by-side with the orthogonal there are oblique fracture sets.

In the event of faults, the description of fracturing becomes more complicated, and if there is no information on fault length and amplitude (amount of displacement), this problem actually becomes insoluble. With fault parameters known, one may judge joint feathering. Fractures at the fault give rise to a belt characterized by a fixed position with respect to the direction of limb displacement. They are closed in the zones of tectonic compression and hence provide only minor water influxes. In the tension zones fractures are open at least to depths of a few tens and hundreds of metres. In competent rocks and at faults, fractures are always wider than they are in incompetent rocks.

The description of fracturing at depth can be undertaken to a high degree of accuracy by studying fractures in shallow openings and even in natural outcrops which have always been transformed by weathering. The quantitative description can be based on the following factors. The arrangement of major fracture sets with respect to layers, fold axial surfaces and fault planes is known not to be depth-dependent. Should the depth position of layers and faults be known, say from drilling and geophysical methods, the trend of major fracture sets and the order of fracture spacing may be determined since fracture spacing varies little with depth and can be assessed in fresh outcrops or shallow (10–30 m) openings.

It is also possible to forecast the form of a fracture at depth on the basis of measurements taken at the surface. Fracture width, a parameter essential to the determination of rock mass permeability and deformability cannot so far be forecast quantitatively.

There is no reliable evidence as to the possibility of predicting average fracture length at depth. The latter, much like fracture width, presumably varies little with depth, a proof of which is the functional relationship between fracture length and width. Thus the order of average fracture length at depth can be predicted on the basis of fracture length measurements made on the surface. Yet in order to determine how deep a particular fracture has penetrated into the rock mass, an opening has to be sunk along the fracture in question.

In regionally metamorphosed sequences the rock fracture patterns are closely related to bedding and schistosity.

In igneous rocks and especially in exo- and endocontact zones of magmatic rock bodies, the rock fracture patterns are more complex since there is no bedding such as that responsible for the evolution of fractures in sedimentary rocks. Even a simple magmatic rock body (a basaltic flow or a granite batholith) features a varied fracture system which is determined by stresses whose orientation changes with time and strength of rock upon cooling, as well as by different cooling rates. It is possible qualitatively to forecast fracturing of magmatic rock bodies on the basis of data on intrusion, its phases, the form of an intrusive or effusive body, arrangement of faults, and the character of modern stresses in a rock mass. Since the above information can be obtained only from detailed investigations, interpolation rather than extrapolation is necessary in the event of intrusive rock masses.

Yet there are certain fairly reliable rules governing fracture patterns in igneous rocks which can be used in forecasting. In zones of crustal tension, fractures in magmatic rock bodies are open to a depth of a few tens and hundreds of metres due to high rock strength. Concentrated inflows into tunnels from fault zones are, therefore, possible. As is well known, effusive rocks, in particular basalts, occasionally feature high porosity and large fracture opening. It is different with abyssal rocks where fractures are mostly healed. Therefore, water permeability of basalts may be high and more uniform than that of abyssal rocks. Granites are known to have fractures extending for several tens of metres, which leads to great deformations. Deformations are favoured by intense mylonitization and kaolinization of granites in fault zones.

Thus, the established rules governing rock fracture patterns can be used in describing rock masses in unstudied areas and in constructing hydrogeomechanical models essential to the calculation of many rock mass parameters in engineering problems.

References

Soviet Authors

(Works by Soviet authors are listed here. For works by authors worldwide see pp.263–266.)

Anon. (1964) Materials on tectonic terminology, Novosibirsk, Part 3, 255 pp. *Proc. Institute of Geology and Geophysics of the Siberian Department of the USSR Academy of Sciences,* issue 34 (in Russian)

Anon. (1975) Studies of the relationship between seismic dislocation and earthquake parameters with orthogonal regression equations. *Proc PNIIIS,* Moscow, issue 39, pp.21–45 (in Russian)

Anon. (1977) *Forms of Geological Bodies,* Glossary, Moscow, Nedra Publishers. 245 pp. (in Russian)

Anon. (1980) *Seismic Regionalization in Northern Mongolia,* Moscow, Nauka Publishers. 179 pp. (in Russian)

Anon. (1980) *Recommendations for Studying Morphology of Rock Fracture Surfaces in the Field and Lab Conditions.* Leningrad VNII9, 88 pp. (in Russian)

Avershin, S. G., V.N. Mosinets and G. P. Cherepanov (1972) On the nature of rock bursts. DAN SSSR, 204 (3), 569–571 (in Russian)

Alekseyev, G. V. (1978) Engineering–Geological Features of Old Intrusive Rock Masses of Basic and Ultrabasic Composition. Synopsis of Thesis for Candidate's (Geology and Mineralogy) Degree, Moscow, MGU, 25 pp. (in Russian)

Batoyan, V. V., and V. N. Korolyov (1976) Hydrochemical anomalies over disjunctive dislocations in the Caspian sea water area. In: *Materials of the 3rd Scientific Conference of Post-Graduate Students and Young Scientists, Ser. Hydrogeology,* Moscow, MGU, pp.36–44 (in Russian)

Batugin, S. A. (1966) Statistical studies of fracture distribution in a rock mass. In: *Rock Mechanics Problems,* Alma-Ata, Nauka Publishers, pp.41–49 (in Russian)

Belikov, B. P. (1964) Elastic constants of rock-forming minerals and their effect on rock elasticity. In: *Physico-Mechanical Properties of Rocks,* Moscow, Nauka Publishers, pp.118–132 (in Russian)

Belitsky, A. A. (1949) *On the Problem of Cleavage Crack Mechanism.* Moscow, AN SSSR Publishers, 49 pp. (in Russian)

Belousov, V. V. (1962) *Basic Problems in Geotectonics.* McGraw-Hill

Belyi, L. D. (1970) On genetic classification of rock fracturing and the position held by seismic fractures in classification system. In: *Seismic Microregionalization of Makhachkala,* Makhachkala, pp.69–87 (in Russian)

Belyi, L. D. (1974) Classification of residual seismic deformations and seismogravitational displacements. *Proc. Moscow Civil Engineering Institute,* col. 111, pp.21–30 (in Russian)

Belyi, L. D. and V. V. Popov (1975) *Engineering Geology.* Moscow, Stroyizdat, 312 pp. (in Russian)

Bogachenko, N. N. (1971) *On the Distribution of Fault Amplitude Exemplified by the Seleznyovskaya Syncline.* Donbass Research Laboratory, USSR Ministry of Geology, issue 3, pp.29–38 (in Russian)

Bogdanov, A. A. (1947) Relationship between cleavage intensity and bed thickness. *Sov. Geology*, No.6, pp.32–41 (in Russian)

Bogdanovich, K. I. (1896) Materials on geology and minerals in the Irkutsk Province. In: *Geological Studies and Exploration in the Siberian Railway Area.* Moscow, issue 2, 284 pp. (in Russian)

Bolotina, N. M. (1970) The engineering–geological environment, Ust-Ilim Hydro. *Hydraulic Project Construction*, No.8, pp.8–11 (in Russian)

Bondarik, G. K. (1959) Release fractures in river valleys, *Exploration and Preservation of the Earth's Interior*, No.10, pp.42–46 (in Russian)

Budko, V. M. (1958) On methods for determining the relative displacement of fault limbs. *Higher School Papers, Mining*, No.3, pp.96–102 (in Russian)

Bukrinsky, V. A. (1970) Studies of the relationship between rock fracturing and offset structures. In: *Problems of Engineering Geology*, Moscow, MGU Publishers, pp.327–332 (in Russian)

Cherepanov, G. P. (1974) *Brittle Failure Mechanics.* Moscow, Nauka Publishers, 640 pp. (in Russian)

Cherepanov, G. P. (1966) On fracture evolution in compressed bodies. *PMM*, 30, pp.83–93 (in Russian)

Chernyshev, S. N. (1965) Exogenic deformation of traps in the Angara Valley. *Proc. of Higher Educational Establishments, Geology and Exploration*, No.12, pp.78–85 (in Russian)

Chernyshev, S. N. (1974) Equations of the relationship between earthquake intensity and seismic dislocation parameters. *BULL. MOIP, Dep. Geol.*, No.1, pp.160 (in Russian)

Chernyshev, S. N. (1979) *Water Movement in Fracture Sets.* Moscow, Nedra Publishers, 140 pp. (in Russian)

Chernyshev, S. N. (1984) *Rock Fracturing and its Effect on Slope Stability.* Moscow, Nedra Publishers (in Russian)

Chernyshev, S. N. (1972) Estimation of the permeability of jointed rocks in a massif. In: *Percolation through Fissured Rock, Proc. Symp. Int. Ass. Eng. Geol.*, Stuttgart, pp.1–11

Chernyshev, S. N. (1976) Standsicherheit einer Felsbösehung mit Klüften begrenzter Länge. *Proc. VI Inter. Conf. Soil Mech. Found. Engineering*, Wien, Vol.1, pp.6–17

Dedova, Ye. V. (1967) Residual deformations of strong earthquakes. In: *Earth Dam Vibrations*, Moscow, Nauka Publishers, pp.114–122 (in Russian)

Dostovalov, B. N. (1959) Some rules governing thermal and diagenetic rock fracturing and the formation of polygonal jointing and textures. In: *Proc. 2nd Conference on Groundwater and Engineering Geology of Eastern Siberia*, Irkutsk Publishers, issue 2, pp.33–43 (in Russian)

Eidelman, S. Ya. (1968) *Field Studies of the Bratsk Hydro Dam*, Leningrad. Energia Publishers, 174 pp. (in Russian)

Famitsyn, B. I. and V. S. Fedorenko (1970) Stereophotogrammetric methods for the quantitative description of rock fractures in slopes of deep foundation pits, quarries and underground mine workings. In: *Aspects of Engineering Geology in Designing, Building and Operating Underground Structures, Mines and Quarries*, Leningrad, USSR Geogr. Soc., issue 2, pp.135–138 (in Russian)

Florensov, N. A. (1960) Neotectonics and seismicity of the Mongolo–Baikalian Highland. *Geology and Geophysics*, No.1, pp.24–36 (in Russian)

Geptner, T. M. (1970) Modelling of shear fractures under conditions of large deformation. *MGU Bull., Ser. 4, Geology*, No.4, pp.81–89 (in Russian)

Gershling, B. M., Yu. N. Litov and A. G. Lopushnyak (1966) Studies into the process of healing individual fractures in clay rocks under water seepage. In: *Hydrogeology and Preservation of Natural Resources when Tapping Salt Deposits*, Leningrad, VNIIG, pp.77–86 (in Russian)

Golodkovskaya, G. A. and L. V. Shaumian (1975) The effect of tectonic fractures on rock mass state and properties. In: *The Effect of Geologic Factors on Rock Mass State and Properties*, Apatity, pp.53–60 (in Russian)

Goncharov, M. A. (1963) On the relationship between cleavage and folding. *MOIP Bull., Dep. Geol.*, 38(4), pp.22–41 (in Russian)

Gorbushina, L. V. and Yu. S. Ryaboshtan (1974) Mapping of zones of recent tectonic movements using radiometry. *Proc. Higher Educational Establishments, Geology and Exploration*, No.6, pp.176–178 (in Russian)

Grigorian, S. S. (1987) *Quantitative Theory of Geocryological Forecasting*. Moscow, MGU Publishers, 296 pp. (in Russian)

Grinbaum, I. I. (1975) *Flow Rate Metering of Hydrogeological and Engineering–Geological Wells*. Moscow, Nedra Publishers, 271 pp. (in Russian)

Gulamov, M. R. (1975) Genesis of fractures normal to the layer in the carbonate fold unit, Maly Karatau Ridge. *Proc. AN KazSSR, Ser. Geol.*, No.2, pp.62–70 (in Russian)

Gzovsky, M. V. (1975) *Fundamentals of Tectonophysics*. Moscow, Nauka Publishers, 525 pp. (in Russian)

Hydrodynamic and Physico-Chemical Rock Properties, N. N. Verigin (ed.) (1977) Moscow, Nedra Publishers, 296 pp. (in Russian)

Ivannikova, O. V. (1988) Determination of the permeability coefficient in a fractured rock mass by calculations using the tubular model. *Power Engineering*, No.8, pp.69–71 (in Russian)

Ivanova, N. B. and S. N. Chernyshev (1974) Computer processing of large-scale measurements of fracturing parameters. *Prc. PNIIIS*, issue 26, pp.186–193 (in Russian)

Ilyin, N. I. and S. N. Chernyshev (1975) Experimental diffusion of a radioactive tracer in an individual fracture. In: *Radioisotopic Methods in Hydrogeology*. Kiev, Naukova Dumka Publishers, pp.122–125 (in Russian)

Ilyin, N. I., S. N. Chernyshev, Ye. S. Dzektser and V. S. Zilberg (1971) *Assessment of Accuracy of Determining Rock Permeability*. Moscow, Nauka Publishers, pp.116–133 (in Russian)

Kalacheva, V. N. (1970) Some data on tectonic fracturing of rocks at depth. *Proc. VNIGRI*, issue 290, pp.106–121 (in Russian)

Kandaurov, I. I. (1966) *Mechanics of Granular Media and its Application to Construction*. Leningrad, Moscow, Stroyizdat Publishers, pp.5–160 (in Russian)

Karpyshev, Ye. S. (1978) *Manual for Determining Water Permeability of Rocks by Water Injection into Holes*. Moscow, Energia Publishers, 44 pp. (in Russian)

Kachanov, L. M. (1974) *Failure Mechanics Principles*. Moscow, Nauka Publishers, 238 pp. (in Russian)

Kirillova, N. V. (1949) Certain aspects of the folding mechanism. *Proc. GEOFIAN*, No. 6, pp.3–91 (in Russian)

Knoring, A. V. (1969) *Mathematical Methods in Studying the Mechanism of Tectonic Fracturing*. Leningrad, Nedra Publishers, pp.3–84 (in Russian)

Kolichko A. V. (1966) Experience in assessing the block size of fractured rock masses. *Proc. Gidroproekt Institute*, issue 14, pp.122–128 (in Russian)

Kolichko, A. V. (1976) Assessment of optimum parameters of rock slopes in fractured rocks. In: *Rock Mass Engineering Geology*. Moscow, Nauka Publishers, pp. 122–126 (in Russian)

Kolichko, A. V. (1981) On the possibility of predicting the amount of displacement on the basis of tectonic faults. *Proc. Gidroproekt Institute*, issue 76, pp. 24–30 (in Russian)

Kolichko, A. V. and M. V. Rats (1966) Fractures in sedimentary rocks in the Toktogulskaya Hydro area and their effect on engineering–geological environment. *Proc. Gidroproekt Institute*, issue 14, pp.104–122 (in Russian)

Kolichko, A. V. and S. N. Chernyshev (1972) On the methods for the experimental injection into holes and filling them with water. In: *Design and Making of Antipercolation Devices in High Dam Foundations*, Moscow, Stroyizdat Publishers, pp.20–26 (in Russian)

Kolmogorov, A. N. (1941) Lognormal law of crushed particle distribution, *DAN SSSR*, 2, 542–546 (in Russian)

Kolomensky, N. V. (1952) *Methodological Guidance for Studying Rock Weathering in Engineering Geology*, Moscow, Gosgeoltekhizdat, 70 pp. (in Russian)

Konyarova, L. P. and L. I. Neustadt (1963) Engineering–geological conditions when building on igneous rocks. In: *Geology and Dams*. Moscow, Leningrad, Gosenergoizdat, vol. 3, pp.9–73 (in Russian)

Korolyov, A. V. (1962) *Ore Field and Deposit Patterns*. Tashkent, Secondary and Higher School Publishers, 148 pp. (in Russian)

Koronovsky, N. V. (1968) Columnar jointing in igneous rocks. *MGU Bull.*, no. 3, pp.91–104 (in Russian)

Kostrov, B. V. (1975) *Mechanism of Tectonic Earthquake Focus*. Moscow, Nauka Publishers, 167 pp. (in Russian)

Kosygin, Yu. A. (1974) *Basic Principles of Tectonics*. Moscow, Nedra Publishers, 213 pp. (in Russian)

Kosygin, Yu. A., I. V. Luchitsky and Yu. A. Rozanov, Experiments on gypsum deformation and their geological significance. *MOIP Bull.*, Dep. Geol., No.2 pp.13–19 (in Russian)

Krasilova, N. S. (1979) Analysis of rock fracturing in small-scale engineering–geological surveys. *Engineering Geology*, No.4, pp.38–46 (in Russian)

Kreiter, V. M. (1956) *Ore Field and Deposit Patterns*. Moscow, GONTI, 272 pp. (in Russian)

Kriger, N. I. (1951) *Fracturing and Methods of Studying Fractures in Hydrogeological Surveys*. Moscow, Metallurgizdat, 152 pp. (in Russian)

Kriger, N. I. and B. P. Preobrazhensky (1953) On the carrying capacity of rocks and rock slope stability. In: *Materials on Engineering Geology*. Moscow, Metallurgizdat, issue 4, pp.65–96 (in Russian)

Kushnarev, I. P. and L. N. Lukin (1960) On the studies of fracture tectonics. In: *Problems of Tectonophysics*. Moscow, Gosgeolizdat, pp.99–119 (in Russian)

Lebedev, A. P. (1955) The trap formation in the central part of the Tunguska Basin, *Proc. GIN, Ser. Petrogr.*, issue 161, 196 pp. (in Russian)

Levinson-Lessing, F. Yu (1940) Petrography. Moscow, Gosgeolizdat, 524 pp. (in Russian)

Lomize, G. M. (1951) *Fractured Rock Permeability*. Moscow, Gosenergoizdat, 127 pp. (in Russian)

Lomtadze, V. D. (1970) *Engineering Geology, Engineering Petrology*. Leningrad, Nedra Publishers, 527 pp. (in Russian)

Lukin, L. I., I. P. Kushnarev and V. F. Chernyshev (1955) On the recurrence of trends of heterochronous fracture systems. *Proc. Institute of Geological Science of the USSR Academy of Sciences*, issue 162, pp.25–35 (in Russian)

Lykoshin, A. G. (1953) Release fractures. *MOIP Bull.*, 28, 4, 53–70 (in Russian)

Lyakhovitsky, F. M., I. P. Morgun and L. G. Chertkov (1961) Experience in mapping fractured zones with the seismic refraction method when making engineering–geological studies under the conditions of a large city. *Engineering Geology*, No. 5, pp.102–106 (in Russian)

Mazanik, V. N. and V. N. Makarov (1970) Rock fractures in the eastern ore cluster in the Pechenga Area and their effect on mining operations. In: *Increasing Open-Pit Mining Efficiency*. Leningrad, Nauka Publishers, pp.84–92 (in Russian)

Manual for Calculating Rock Mass Permeability Coefficient on the Basis of Fracture Parameters, Moscow, Stroyizdat Publishers, 1979, 60 pp. (in Russian)

Anon. (1964) Materials on tectonic terminology, Novosibirsk, Part 3, 255 pp. *Proc. Institute of Geology and Geophysics of the Siberian Department of the USSR Academy of Sciences*, issue 34 (in Russian)

Smekhov, Ye. M. (ed) 1969 *Methods for Studying Rock Fracturing and Fractured Oil and Gas Reservoirs*. Leningrad, Nedra Publishers, 129 pp. (*Proc. VNIGRI*, issue 276) (in Russian)

Mindel, I. G. and A. P. Golubkov (1974) On identification of near–surface dislocation using diffracted wave data. In: *Seismic Regionalization and Engineering-Geological Survey*. *Proc. PNIIIS*, issue 30, pp.88–98 (in Russian)

Mironenko, V. A. and V. M. Shestakov (1974) *Fundamentals of Hydrogeomechanics*. Moscow, Nedra Publishers, 295 pp. (in Russian)

Mikhailov, A. Ye (1956) *Field Techniques for Studying Rock Fractures.* Moscow, Gosgeoltek-hizdat, 132 pp. (in Russian)

Myachkin V. I. (1948) *Earthquake Preparation Processes.* Moscow, Nauka Publishers, 232 pp. (in Russian)

Novikova, A. S. (1951) On rock fractures in the eastern part of the Russian platform. *Proc. AN SSSR, Ser. Geol.,* No.5, pp.68–85 (in Russian)

Ovchinnikov, A. M. (1938) On the methods for studying fractures, *Exploration of the Earth's Interior,* nos 4–5, pp.32–41 (in Russian)

Pavlova, N. V. (1970) *Fracturing and Rock Failure,* Moscow, Nauka Publishers, pp.7–15 (in Russian)

Pek, A. V. (1939) *Fracture Tectonics and Structural Analysis.* Moscow, Leningrad, AN SSSR Publishers, 152 pp. (in Russian)

Pashkin Ye. M. (1981) *Engineering–Geological Studies when Building Tunnels.* Moscow, Nedra Publishers, 135 pp. (in Russian)

Peive, A. V. (1956) The principle of inheritance in tectonics. *Proc. AN SSSR, Ser. Geol.,* No.6, pp.11–19 (in Russian)

Peive, A. V. (1965) Horizontal crustal movements and the inheritance principle. *Geotectonics,* No. 1, pp.30–37 (in Russian)

Pogrebisky, M. I., M. V. Rats and S. N. Chernyshev (1971) On relationship between fracture spacing and fault distance. *DAN SSSR,* **201,** 4, pp.927–930 (in Russian)

Pogrebisky, M. I. and S. N. Chernyshev (1974) Feather joints in seismic strike slip faults. *DAN SSSR,* **218,** 5, pp.1171–1174 (in Russian)

Polkanov, A. A. (1956) Genetic system of cratogene platform intrusions. *Proc. AN SSSR, Ser. Geol.,* No. 6, pp.5–29 (in Russian)

Priklonsky, V. A. (1952) *Soil Science.* Moscow, Gosgeoltekhizdat, 1952, vol. II, 371 pp. (in Russian)

Protodyakonov, M. M. and Ye. S. Chirkov (1964) *Rock Mass Fracturing and Strength.* Moscow, Nauka Publishers, 77 pp. (in Russian)

Prochukhan, D. P. (1964) Release fractures in rock beds of high dams. *Soviet Geology,* No. 7, pp.74–83 (in Russian)

Rats, M. V. and S. N. Chernyshev (1967) Statistical aspect of the problem of permeability of jointed rocks. In: *Hydrology of Fractured rocks, Vol.1. Proc. Dobrovnik Symposium of the Intern. Ass. of Scientific Hydrology, Oct. 1965,* printed by Centerick, Louvain (Belgium), pp.227–236

Rats, M. V. and S. N. Chernyshev (1970) *Fracturing and Properties of Fractured Rocks.* Moscow, Nedra Publishers, 160 pp. (in Russian)

Rebrik, B. M. and S. N. Chernyshev (1968) Relationship between measuring accuracy and heterogeneity of the medium and design requirements. In: *Mathematical Methods in Engineering Geology.* Moscow, MOIP, pp.121–125 (in Russian)

Rock fracture parameters: measuring accuracy. *Proc. PNIIIS,* issue 39, pp.80–99 (in Russian)

Romm, Ye. S. (1966) *Percolation Properties of Fractured Rocks.* Moscow, Nedra Publishers, 282 pp. (in Russian)

Ruppeneit, K. V. (1975) *Fractured Rock Mass Deformability.* Moscow, Nedra Publishers, 222 pp. (in Russian)

(1975) *Reference Book on Physical Rock Properties,* (eds) Melnikov, N. V., V. V. Rzhevsky and M. M. Protodyakonov. Moscow, Nedra Publishers, 277 pp. (in Russian)

Medvedev, S. V. (ed) (1971) *Seismic Regionalization in Ulan-Bator.* Moscow, Nauka Publishers, 285 pp. (in Russian)

Sergeyev, Ye. M. (1974) On the future of engineering geology. *MGU Bull.,* No. 1, pp.7–15 (in Russian)

Shafranovsky, I. I. (1968) *Symmetry in Nature.* Moscow, Nedra Publishers, 184 pp. (in Russian)

Shestakov, V. M. (1961) On the theory of sorption dynamics under percolation in granular media. *ZhFKh,* **35,** 10, 2358–2362 (in Russian)

Stini, I. and D. I. Mushketov (1934) *Technical Geology.* Moscow, Gosneftizdat, 256 pp. (in

Russian)

Shultz, S. S. (1971) Planetary fractures. *Geotectonics*, No.3, pp.18–34 (in Russian)

Skaraytin, V. D. (1965) Classification of fractures and aerial photography and some other techniques employed for rock fracture studies. *Proc. 2nd All-Union Conference on Fracture Oil and Gas Reservoirs*. Moscow, Nedra Publishers, pp.5–14 (in Russian)

Skvortsov, G. G. and V. V. Fromm (1970) *Engineering–Geological Studies of the Deep Levels of Mineral Deposits in Exploration*. Moscow, Nedra Publishers, 110 pp. (in Russian)

Solonenko, V. P. (1970) Determination of epicentral earthquake zones based on geological evidence. *Proc. AN SSSR, Ser. Geol.*, No. 11, pp.58–74 (in Russian)

Solonenko, V. P. (1973) Paleoseismogeology. *Proc. AN SSSR, Physics of the Earth*, No. 9, pp.72–85 (in Russian)

Sokolov, D. S. (1968) *Major Karstification Conditions*. Moscow, Gosgeoltekhizdat, 322 pp. (in Russian)

Sokolov, N. I. (1961) On the types of dislocation of fractured rocks in slopes. *Proc. Lab. Hydrogeol. Probl.*, issue 35, pp.25–38 (in Russian)

Sorokin, A. A., V. N. Badukhin and F. V. Shpakovsky (1972) Method for aerial determination of rock fracturing. In: *Theses of Papers of the 4th Gidroproekt Institute Explorers' Conference on Sharing Experience in the Construction of Hydraulic Projects*, Eng. Geology Section, issue 2, pp.48–50 (in Russian)

Stoyanov, S. S. (1977) *Mechanism of Fault Zone Formation*. Moscow, Nedra Publishers, 112 pp. (in Russian)

Tarkhov, A. G. (1940) On anisotropy of elastic rock properties, *VSEGEI Materials, Gen. Ser.*, col. 5, pp.209–222 (in Russian)

Ter-Stepanian, G. I. and A. P. Arakelian (1975) On superimposition of fracturing in basalt lava flows. In: *The Effect of Geological Factors on Rock Mass State and Properties*, Apatity, pp.141–145 (in Russian)

Tetyayev, M. M. (1940) Tectonics of veined ore fields, *Sov. Geology*, nos. 8, 9, pp.16–24 (in Russian)

Tizdel, R. R. (1962) The Angara Bratsk Dam. In: Geology and Dams, Moscow, *Gosenergoizdat*, **2**, pp.22–45 (in Russian)

Tizdel, R. R. (1963) Bratsk hydro: rock bed settlement. *Hydraulic Project Construction*, No. 9, pp.18–19 (in Russian)

Tkachuk, E. I. (1967) Application of Correlation Analysis Methods to Engineering–Geological Investigations of Igneous Rocks. Synopsis of Thesis for Candidate's (Geology and Mineralogy) Degree, Leningrad. Leningrad Mining Institute, 20 pp. (in Russian)

Tolokonnikov, I. S. (1966) Lavas in Armenia: fracturing and permeability. *Proc. Gidroproekt Institute*, issue 14, pp.150–179 (in Russian)

Ukhov, S. B. (1975) *Rock Beds of Hydraulic Structures*. Moscow, Energia Publishers, 262 pp. (in Russian)

Ventsel, Ye. S. (1964) *Theory of Probability*. Moscow, Fizmatgiz, 264 pp. (in Russian)

Vistelius, A. B. (1958) *Structural Diagrams*. Moscow, Leningrad, AN SSSR Publishers, 165 pp. (in Russian)

Voznesensky, A. V. (1962) *Investigation of the Khangai Earthquake (1905) Region in Northern Mongolia*. Leningrad, USSR Geogr. Society, 70 pp. (in Russian)

Yanshin A. L. (1951) A. D. Arkhangelsky's view of the tectonic nature of the south-eastern fringe zone of the Russian platform and present-day theories on the subject. In: *Problems of USSR Lithology and Stratigraphy*, Moscow, AN SSSR Publishers, pp.253–327 (in Russian)

Yegorov, A. Ya (1981) Application of CO_2 survey to tracing local faults in central Moldavia. *VIEMS Express Bull., Hydrogeology and Eng. Geology*, Moscow, issue 6, pp.6–10 (in Russian)

Yeliseyev, N. A. (1953) *Structural Petrology*. Leningrad, LGU Publishers, 310 pp. (in Russian)

Zhilenkov, V. N. (1975) *Manual for the Determination of Percolation and Suffosive Properties of Rock Beds of Hydraulic Structures.* Leningrad, Energia Publishers, 75 pp. (in Russian)

Zolotarev, G. S. (1962) The weathering crust in Archaean rocks in Ceylon and its engineering–geological significance, *Proc. Higher Educational Establishments, Geology and Exploration,* No. 2, pp.61–70 (in Russian)

Authors worldwide

(Works listed here are those by authors worldwide, excepting Soviet authors. For works by Soviet authors see pp.257–263.)

Anon. (1970). The logging of rock cores for engineering purposes. *Quarterly Journal Engineering Geology,* **3**, 1–25

Anon. (1972) The preparation of maps and plans in terms of engineering geology. *Quarterly Journal Engineering Geology,* **5**, 295–382

Anon. (1976) *Engineering Geological Maps: A guide to their preparation.* The Unesco Press, Paris, pp.79

Anon. (1977) The description of rock masses for engineering purposes. *Quarterly Journal Engineering Geology,* **10**, 355–388

Anon. (1981a) Code of Practice for Site Investigations (Formerly CP2001), BS5930:1981. British Standards Institution, London, pp.147

Anon. (1981b) Rock and soil description and classification for engineering geological mapping. *Report by the IAEG Commission on Engineering Geological Mapping. Bulletin International Association Engineering Geology,* No. 24, pp.227–234

Aydin, A. and J. M. Degraff (1988) Evolution of polygonal fracture patterns in lava flows. *Science,* **239**, 471–476

Balk, R. (1937) *Structural Behaviour of Igneous Rocks.* Geological Society America, Memoir 5

Barr, M. V. and G. Hocking (1977) Borehole structural logging employing a pneumatically inflatable impression packer. *Proc. Symposium Exploration for Rock Engineering Johannesburg.* Balkema, Rotterdam, pp.29–34

Barton, C. C. (1983) Systematic Jointing in the Cardium Sandstone Along the Bow River, Alberta, Canada (PhD Thesis). New Haven, Connecticut, Yale University, pp.301

Barton, N. (1976) The shear strength of rock and rock joints. *International Journal Rock Mechanics Mining Science and Geomechanics Abstracts,* **13**, 255–279

Barton, N. (1987) Discontinuities. In Bell, F. G. (ed.) *Ground Engineer's Reference Book.* Butterworths, London, 5/1–5/15

Barton, N., R. Lien and J. Lunde (1974) *Engineering Classification of Rock Masses for the Design of Tunnel Support.* Norwegian Geotechnical Institute. Publication 106, pp.1–48

Bates, R. L. and J. A. Jackson (ed.) (1980) *Glossary of Geology.* American Geological Institute, 2nd edn, pp.751

Bell, F. G. (ed) (1987) *Ground Engineer's Reference Book.* Butterworths, London (59 Chapters by specialist contributors)

Bieniawski, Z. T. (1976) Rock mass classification in rock engineering. *Proc. Symposium Exploration for Rock Engineering, Johannesburg,* Vol. 1, pp.97–106

Bieniawski, Z. T. (1979) The geomechanics classification in rock engineering applications. *Proc. 4th International Congress on Rock Mechanics,* International Society Rock Mechanics. Montreux Switzerland, Vol. 2, pp.41–48

Brown, E. T. (ed.) (1981) Rock characterization testing and monitoring. *International Society Rock Mechanics Suggested Methods.* Pergamon Press, pp.211

Burger, H. R. and M. D. Thompson (1970) Fracture analysis of the Carmichael Peak Anticline, Madison County, Montana. *Bull. Geological Society America,* **81**, 1831–1835

Cailleux, A. (1958) Etude quantitative de failles. *Revue Geomorphologie Dynamique*, **9**, 129–146

Cloos, H. (1923) Das Batholithenproblem. *Fortschritte der Geologie und Palaeontologie, Berlin*, **1**

Dearman, W. R. (1974) Weathering classification in the characterization of rock for engineering purposes. *Bull. International Association Engineering Geology*, No.9, 33–42

Dearman, W. R. (1976) Weathering classification in the characterization of rock: a revision. *Bull. International Association Engineering Geology*, No. 13, pp.123–127

Deere, D. U. (1968) Geological considerations. In *Rock Mechanics in Engineering Practice*, M. G. Stagg and O. C. Zienkiewiez (eds.). Wiley, London, pp.1–19

Dershowitz, W. S. and H. H. Einstein (1987) Three-dimensional flow modelling in jointed rock masses. *Proc. 6th Congress Rock Mechanics*, Montreal

Dyer, J. R. (1983) Jointing in Sandstones, Arches National Park, Utah. (PhD Thesis). Stanford, California, Stanford University, pp.202

Elsworth, D. and A. R. Piggot (1987) Physical and numerical analogues to fractured media flow. *Proc. 6th Congress Rock Mechanics*, Montreal

Fecker, E. and N. Rengers (1971) Measurement of large-scale roughnesses of rock planes by means of a profilograph and geological compass. *Rock Fracture. Proc. International Symposium Rock Mechanics*, Nancy, Paper I.18

Farran, J. and B. Thenoz (1965) L'alterabilite des roches, des fractures, sa prevision. *Annales del I.T.B.T.P.*, No. 215

Franklin, J. A. (1986) Photo-analysis of rock jointing. *Proc. 39th Canadian Geotechnical Conference*, Ottawa, Canada

Franklin, J. A. (1988) Rock mass characterization using photoanalysis. *International Journal Mining Geological Engineering*, **6**, 97–112

Griffith, A. A. (1921) The phenomenon of rupture and flow in solids. *Philosophical Transactions Royal Society, London*, **A228**, 163–197

Griffith, A. A. (1924) Theory of rupture. *Proc. 1st International Congress Applied Mechanics*, Delft, pp.55–63

Gruneisen, P., G. Hirlemann, P. Janot and M. Ruhland (1973) Analyse de la fracturation naturelle d'une structure plissee: l'anticlinal de la Lance (Drome). *Sci. geol. Bull.*, **26**, 161–186

Hattori, I. and S. Mizutani (1971) Computer simulation of fracturing of layered rock. *Engineering Geology*, **5**, 253–269

Hill, P. A. (1965) Curviplanar and concentric jointing in Jurassic dolerite, Mersey Bluff, Tasmania. *Journal Geology*, **73**, 255–271

Hinds, D. V. (1974) A method of taking an impression of a borehole wall. *Imperial College, London, Rock Mechanics Research Report Number 28*, pp.10

Hoek, E. (1973) Methods for the rapid assessment of the stability of three-dimensional rock slopes. *Quarterly Journal Engineering Geology*, **6**, 243–255

Hoek, E. and J. Bray (1977) *Rock Slope Engineering* (revised 2nd edn). Institution of Mining and Metallurgy, London, pp.402

Hoek, E. and E. T. Brown (1980) *Underground Excavations in Rock*. The Institution of Mining and Metallurgy, London. pp.527

Hudson, A. and S. D. Priest (1979) Discontinuities and rock mass geometry. *International Journal Rock Mechanics Mining Science and Geomechanics Abstracts*, **16**, 339–362

Huitt, J. L. (1956) Fluid flow in simulated fracture. *Journal American Institution Chemical Engineering*, **2**, 259–264

Iida, K. (1959) Earthquake energy and earthquake fault. *Journal Earthquake Science University Nagoya, Japan*, **7**, 99–167

Inglis, C. E. (1913) Stresses in a plate due to the presence of cracks and sharp corners. *Royal Institute of Naval Architects Transactions*, **55**, 219–230

Irwin, G. R. (1958) Fracture. In Flugge, S. (ed), *Encyclopedia of Physics*. Berlin, Springer-Verlag, pp.551–590

Jaeger, C. (1972) *Rock Mechanics and Engineering*. Cambridge University Press, pp.417

James, A. N. and A. R. R. Luptok (1978) Gypsum and anhydrite in foundations of hydraulic structures. *Geotechnique*, **28**, 249–282

John, K. W. (1968) Graphical stability analysis of slopes in jointed rocks. *Proc. American Society Civil Engineers*, Journal Soil Mechanics and Foundation Engineering, **94**, SM2, 497–526

Jueguang Jiang (1986) Stereographic projection analysis for exposed conditions of slide-falling blocks in jointed rockmass. *Proc. International Symposium Engineering in Complex Rock Formations*, Beijing, pp.816–822

Ketin, J. and F. Roesli (1953) Makroseismische Untersuchungen uber das nordwest Anatolische Beben von 18 Marz 1953. *Eclogae Geologicae Helvetiae*, **47**, 50–64

Kieslinger, A. (1958) Restspannung und Entspannung im Gestein. *Geologie und Bauwesen*, **24**, 95–112

Kranz, R. L (1983) Microcracks in rocks: a review. *Tectonophysics*, **100**, 449–480

Louis, C. (1968) Etude des ecoulements d'eau dans les roches fissurees et de leurs influences sur las stabilite das massifs rocheux. *Bull. Direction etude et recherches.*, **A** (3), 5–132

Macovec, F. (1962) Das Ausmass der Felsauflockerung bei Sprengarbeitern. *Geologie und Bauwesen*, **28**, 58–62

McQuillan, H. (1974) Fracture patterns in the Kuh-e Asmari Anticline, South-west Iran. *Bull. American Association of Petroleum Engineering*, **58**, 236–246

Moore, J. F. (1974) Mapping major joints in the Lower Oxford Clay using terrestrial photogrammetry. *Quarterly Journal Engineering Geology*, **7**, 57–67

Müller, K. E. (1974) Zur Definition des Durchtrennungsgrades. *Rock Mechanics (Wien)*, Supplement, **3**, 17–19

Müller, L. (1961) Uber die Enstehung oberflachenparalleler Klufte. *Geologie und Bauwesen*, **27**, 146–152

Muller, L. (1963) *Der Felsbau*. Ferdinand Enke Verlag Stuttgart, pp.624

Nickelsen, R. P. and V. N. D. Hough (1967) Jointing in the Appalachian Plateau of Pennsylvania. *Bull. Geological Society of America*, **78** 609–629

Patton, F. D. (1966) Multiple modes of shear failure in rock. *Proc. 1st International Congress Rock Mechanics*, Lisbon, **1**, 509–514

Phillips, F. C. (1971) *The Use of Stereographic Projection in Structural Geology*. Edward Arnold, London, (3rd edn paperback) pp.90

Piteau, D. R. (1973) Characterizing and extrapolating rock joint properties in engineering practice. *Rock Mechanics*. Springer-Verlag, Supplement 2, 5–31

Pollard, D. D. and A. Aydin (1988) Progress in understanding jointing over the past century. *Geological Society of America Special Bulletin*, **100**, 1181–1204

Pollard, D. D. and P. Segall (1987) Theoretical displacements and stresses near fractures in rock, with applications of faults, joints, veins, dykes and solution surfaces. In B. K. Atkinson (ed), *Fracture Mechanics in Rock*. London, Academic Press, pp. 277–349

Preuss, H. D. (1974) Numerisch-photogrammetrische Kluftmessung. *Rock Mechanics, Wien*, Supplement, **3**, 5–15

Price, N. J. (1966) *Fault and Joint Development in Brittle and Semibrittle Rock*. Pergamon, Oxford

Priest, S. D. and J. A. Hudson (1976) Discontinuity spacings in rock. *International Journal Rock Mechanics Mining Science and Geomechanics Abstracts*, **13**, 135–148

Priest, S. D. and J. A. Hudson (1981) Estimation of discontinuity spacing and trace length using scanline surveys. *International Journal Rock Mechanics Mining Science and Geomechanics Abstracts*, **18**, 183–197

Richey, J. E. (1963) Granite, *Water Power*, **6**, 61–65

Richter, H. O., H. Molek, and F. Reuter (1976) Methodische Probleme bei der Ermitlung strukturgeologischer Parameter im Fels. *Zeitschrift fur angewandte Geologie. Zentrales Geologisches Institut. Berlin. DDR*, **5**, 237–243

Riedel, W. (1929) *Zur Mechanik geologischer Brucherscheinungen*. Centralblatt fur Mineralogie, Geologie und Palaeontologie. Stuttgart. B. 354

Riffaud, C. and Z. Le Pichon (1976) *Expedition 'FAMOUS'. Paris*

Robertson, A. MacG. (1971) The interpretation of geological factors for use in slope theory. *Symposium on Planning Open Pit Mines, Johannesburg*. Balkema, Amsterdam, pp.55–71

Ross-Brown, D. M. and K. B. Atkinson (1972) Terrestrial photogrammetry in open pits: 1 – description and use of the phototheodolite in mine surveying. *Transactions Institution Mining Metallurgy* (Section A. Mining Industry), **81**, A205–A213

Ross-Brown, D. M., E. H. Wickens and J. T. Markland (1973) Terrestrial photogrammetry in open pits: 2 – an aid to geological mapping. *Transactions Institution Mining Metallurgy* (Section A. Mining Industry), **82**, A115–A130

Ruxton, B. R. and L. Berry (1957) Weathering of granite and associated erosional features in Hongkong. *Bull. Geological Society of America*, **68**, 1263–1292

Sander, B. (1930) *Gefugekunde der Gesteine*. Vienna, Julius Springer, pp.352

Savage, W. and D. Varnes (1987) Mechanics of gravitational spreading of steep-sided ridges. *Bull. International Association of Engineering Geology*, No. 35, pp.31–36

Segall, P. and D. D. Pollard (1983a) Nucleation and growth of strike slip faults in granite. *Journal of Geophysical Research*, **88**, 555–568

Segall, P. and D. D. Pollard (1983b) Joint formation in granitic rock of the Sierra Nevada. *Bull. Geological Society of America*, **94**, 563–575

Skempton, A. W. (1966) Some observations on tectonic shear zones. *Proc. 1st Congress International Society Rock Mechanics, Lisbon*, **I**, 329–335

Skipp, B. O. and N. N. Ambraseys (1987) Engineering seismology. In Bell, F. G. (ed) *Ground Engineer's Reference Book*. Butterworths, London, 18/1-18/25

Smally, I. I. (1966) Contraction crack networks in a basalt flow. *Geological Magazine, London*, **103**, 110–113

Snow, D. T. (1968) Rock fracture spacings, openings and porosities. *Proc. American Society Civil Engineers. Journal Soil Mechanics and Foundation Engineering Division*, **94**, 73–91

Spencer, A. B. and P.S. Clabaugh (1967) Computer programs for fabric diagrams. *American Journal of Science*, **265**, 166–172

Tchalenko, J. S. (1970) Similarities between shear zones of different magnitudes. *Bull. Geological Society of America*, **81**, 1625–1640

Terzaghi, K. (1962) Dam foundation on sheeted granite. *Geotechnique*, **12**, 199–208

Terzaghi, R. D. (1965) Sources of error in joint surveys. *Geotechnique*, **15**, 387–304

Tocher, D. (1958) Earthquake energy and ground breakage. *Bulletin of the Seismological Society of America, Baltimore*, **48**, 147–152

van Hise, C. R. (1896) Principles of North American Pre-Cambrian geology. *US Geological Survey 16th Annual Report*, pp.581–874

Vieten, V. (1970) Die Ermittlung tektonischer Gefugedaten aus stereofotogrammetrischen Bruch Wanderaufnahmen. *Clausthaler tekton*, Hefte, **10**, 316–328

Wallis, P. F. and M. S. King (1980) Discontinuity spacings in a crystalline rock. *International Journal Rock Mechanics Mining Science and Geomechanics Abstracts*, **17**, 63–66

Walsh, J. B. (1965) The effects of cracks on the uniaxial compression of rocks. *Journal Geophysical Research, Richmond*, **70**, 381–411

Wilson, R. G. and P. A. Witherspoon (1974) Steady-state flow in rigid networks of fractures. *Water Resources Research*, **10**, 328–335

Witherspoon, P. A. (1986) Flow of groundwater in fractured rocks. *Bull. International Association Engineering Geology*, No. 34, 103–115

Wittke, W. (1970) Three-dimensional percolation of fissured rocks. In *Proc. Symposium on the Theoretical Background to the Planning of Open Pit Mines with Special Reference to Slope Stability*. South African Institute of Mining and Metallurgy, Johannesburg, pp.78–86

Wittke, W. (1984) *Felsmechanik: Grundlagen fur wirtsdfeftliches Bauen im Fels*. Springer-Verlag, Berlin, pp.1050

Wittke, W. and C. Louis (1966) Zur Berechnung des Einflusses der Bergwasserstromung auf die Standsicherheit von Boschnungen und Bauwerken in zerklufterten Fels. *Proc. 1st Congress International Society of Rock Mechanics, Lisbon*, **II**, 6.14

Woodworth, J. B. (1896) On the fracture system of joints, with remarks on certain great fractures. *Proc. Boston Natural History Society*, **27**, 163–183

Index